AA001046

2018 IEEE 8th International Nanoelectronics Conference (INEC 2018)

Kuala Lumpur, Malaysia
3-5 January 2018

IEEE Catalog Number: CFP18625-POD
ISBN: 978-1-5386-4251-1

**Copyright © 2018 by the Institute of Electrical and Electronics Engineers, Inc.
All Rights Reserved**

Copyright and Reprint Permissions: Abstracting is permitted with credit to the source. Libraries are permitted to photocopy beyond the limit of U.S. copyright law for private use of patrons those articles in this volume that carry a code at the bottom of the first page, provided the per-copy fee indicated in the code is paid through Copyright Clearance Center, 222 Rosewood Drive, Danvers, MA 01923.

For other copying, reprint or republication permission, write to IEEE Copyrights Manager, IEEE Service Center, 445 Hoes Lane, Piscataway, NJ 08854. All rights reserved.

*** This is a print representation of what appears in the IEEE Digital Library. Some format issues inherent in the e-media version may also appear in this print version.

IEEE Catalog Number: CFP18625-POD
ISBN (Print-On-Demand):   978-1-5386-4251-1
ISBN (Online):   978-1-5386-4250-4
ISSN: 2159-3523

Additional Copies of This Publication Are Available From:

Curran Associates, Inc
57 Morehouse Lane
Red Hook, NY 12571 USA
Phone: (845) 758-0400
Fax: (845) 758-2633
E-mail: curran@proceedings.com
Web: www.proceedings.com

TABLE OF CONTENTS

RELIABILITY PERSPECTIVE ON THE IOT AND NANOELECTRONICS..1
    C. Tan

VERTICAL NANOWIRE TUNNELING FIELD-EFFECT TRANSISTORS ADOPTING CORE-SHELL STRUCTURE WITH STRAIN EFFECTS 3
    J.-S. Yoon, C.-K. Baek

DEVELOPMENT OF MALODOR MONITORING SYSTEM BASED ON ELECTRONIC NOSE TECHNOLOGY 5
    T. Pobkrut, S. Siyang, T. Thepudom, T. Kerdcharoen

THERMAL CHARACTERIZATION OF POLYCRYSTALLINE DIAMOND USING INFRARED THERMAL IMAGING MEASUREMENT 7
    B. Tay, Q. Kong, L. Bodelot, B. Lebental, D. Misra

A 2DOF MEMS VIBRATIONAL ENERGY HARVESTER..9
    K. Tao, L. Tang, J. Wu, J. Miao

HCI AND NBTI RELIABILITY SIMULATION FOR 45NM CMOS USING ELDO..............................11
    A. Jaafar, N. Soin, S. Hatta

A COMPENSATION METHOD FOR LONG-TERM ZERO BIAS DRIFT OF MEMS BYROSCOPE BASED ON IMPROVED CEEMD AND ELM...13
    H. Gu, X. Liu, B. Zhao, H. Zhou

EVOLUTION OF THE PHYSICS AND STOCHASTICS OF FAILURE IN ULTRA-THIN DIELECTRICS - FROM SIO_2 TO ADVANCED HIGH-K GATE STACKS 15
    N. Raghavan

STRATEGIES IN MICROFLUIDIC SELF-ASSEMBLED NANOPARTICLES....................................17
    C. Iliescu, G. Tresset, M. Ni, C. Marculescu

ELECTRICAL ANALYSIS OF INGAAS-BASED PLANAR AND TRI-GATE NMOSFET WITH S/D RESISTANCE DEPENDENCIES AT DIFFERENT DRAIN BIASES 19
    N. Othman, S. Hatta, N. Soin

FLUXLESS FLIP CHIP INTERCONNECTS FOR MEMS DEVICES FOR HETEROGENEOUS INTEGRATION 21
    T. Lee

AN ENHANCED HIGH-SENSITIVITY MICRO RESONANT THERMOMETER WITH AXIAL STRAIN AMPLIFICATION EFFECT 23
    Q. Shen, J. Yang, J. Xie, S. Ren, W. Yuan

ELECTROMECHANICAL PIEZORESISTIVE SENSING OF GRAPHENE-BASED INTRACRANIAL PRESSURE SENSOR 25
    M. Mohamad, N. Soin, F. Ibrahim

PIEZORESISTIVE EFFECT OF INTERDIGITATED ELECTRODE SPACING GRAPHENE-BASED MEMS INTRACRANIAL PRESSURE SENSOR 27
    S. Rahman, N. Soin, F. Ibrahim

MULTI-LAYER NONCONTACT DISK-SHAPED ELECTROSTATIC MICROGENERATOR 29
    B. Wang, Y. Chen, L. Tang, K. Tao

DISTRIBUTION OF SN IN STRAINED $GE_{1-X}SN_X$ (001): THE EFFECT OF SURFACE PASSIVATION..31
    S. Ong, E. Tok

CONCEPT, METHODOLOGIES AND TOOLS FOR CARBON FOR SENSING DEVICES.................33
    M. Giorcelli, P. Savi, A. Tagliaferro

TUNING ENDOTAXIAL GROWTH OF COSI2 NANOWIRES AND NANODOTS................................35
    B. Ong, E. Tok

AN IMPROVED SOI RESONANT PRESSURE SENSOR USING ATMOSPHERIC PACKAGING 37
    S. Ren, J. Xie, Q. Shen, F. Wang, W. Yuan, J. Zhang

FLEXIBLE PRESSURE AND FORCE SENSING SYSTEM FOR WEARABLE MEDICAL DEVICES 39
    N. Xue, C. Liu, J. Sun, C. Wang

PREPARATION OF MULTI-WALLED CARBON NANOTUBES/POLYDIMETHYLSILOXANE COMPOSITE FOR ELECTRONIC SKIN APPLICATION...41
    C. Chi, X. Sun, T. Li, N. Xue, C. Liu

WATER HARDNESS DETERMINATION USING DISPOSABLE MEMS-BASED ELECTROCHEMICAL SENSOR......43

N. Wang, E. Kanhere, K. Tao, J. Wu, J. Miao, M. Triantafyllou

THE FABRICATION METHOD OF HIGH HEIGHT-TO-DIAMETER RATIO MICRO-WINEGLASS RESONATORS......45

J. Xie, S. Ren, Q. Shen, L. Chen, H. Xie, W. Yuan

IN-PLANE ROTATIONAL TUNING OF POLYMER DIFFRACTION GRATING FOR DIVERSE IMAGING SPECTROSCOPY......47

S. Muttikulangara, M. Baranski, S. Rehman, L. Hu, J. Miao

SENSITIVITY AND Q-FACTOR TRADE-OFF ANALYSIS OF MEMS PRESSURE SENSOR FOR BLADDER IMPLANTS......49

N. Yusof, B. Bais, B. Majlis, N. Soin

EFFECT OF FLUORINE CIRCUMFERENCE OF ZINC-HEXADECAFLUOROPHTHALOCYANINE TOWARDS VOCS DETERMINATION BY USING LOW-COST OPTICAL ELECTRONIC NOSE......51

T. Thepudom, T. Kerdcharoen

ANALYTICAL CAPACITANCE MODEL FOR IN-GA-ZN-O THIN-FILM TRANSISTORS INCLUDING DEGENERATION......53

F. Zhuang, J. Fang, W. Deng, X. Ma, J. Huang

Author Index

Reliability Perspective on the IoT and Nanoelectronics

Cher Ming Tan

Centre for Reliability Sciences & Technology, Chang Gung University, Taoyuan, Taiwan,
Department of Electronic Engineering, Chang Gung University, Taoyuan, Taiwan
Department of Urology, Chang Gung Memorial Hospital, Taoyuan, Taiwan,
Department of Mechanical Engineering, Ming Chi University of Technology, New Taipei City, Taiwan

cmtan@cgu.edu.tw

Abstract - Internet of Things (IoT) is coming strongly and we have no escape to it. While current technology and its development can indeed make IoT a reality, its cost has to be much lower and at the same time its reliability must be very high. Furthermore, the necessity of nano-electronics in IoT presents new degradation mechanisms that need to be studied in detail. Thus, it is a huge challenge in reliability field in anticipation of IoT era. This work presents some critical issues and highlight some works that are being done to meet the coming challenges.

Index Terms — Radiation, electromigration, electromagnetics, new failure mechanisms

INTRODUCTION

Gartner predicts that there will be 20.4 billion connected "things" by 2020, and these "things" will come in all shapes and sizes [1]. Basically, we have no escape to the effect of the IoT.

At present, there is already 1 billion phone calls per day in Western Europe, 247 billion email messages per day, 60 million Twitter messages per day [2]. All these numbers are increasing and will be a lot more with the IoT. Among these messages, some are of critical important and their inability to transmit could result in serious consequences. Thus, the reliability of the interconnectivity is of stringent requirement.

The high data rate transmission and high functionality requires integrated circuits to be faster and highly complex. Nano-devices are essential, and they are interconnected through nano-interconnections. It is well known that high complexity leads to poor reliability unless specific actions are taken with this anticipation. Also, with nano-devices and interconnections, different reliability considerations are needed due to changing degradation mechanisms at nano-scale [3]. In this work, some new critical degradation mechanism will be presented.

RADIATION SENSITIVITY

Nano-scale transistors are more sensitivity to radiation. The soft error rate of 22 nm technology is 7 times higher than its 130 nm counterpart [4], and the critical charge, which is the amount of charge necessary to flip a binary "1" to "0" or vice versa, is also decreasing drastically 24 times from 130 nm to 22 nm [5].

At sea level, neutrons are more prominent among other high energy particles [6]. Although the flux of neutrons is

978-1-5386-4251-1/18 $31.00 © 2018 IEEE

low and insignificant to general electronics, as the semiconductor devices become nano-scale, radiation hardness is no longer for space exploration. Evaluation capability for neutron radiation effect on electronic devices is now established by the author, in view of the need.

Electromigration (EM) and Electromagnetic emission Electromigration in nano-interconnections in VLSI is expected to be more drastic as very tiny void can result in open circuit instead of gradual increase in resistance as shown by Tan et al. [7], renders shorter EM lifetime. Recently, a cost effective method to incorporate graphene onto copper interconnect is developed [8], and the improvement in EM lifetime is found to be 3.33 times [9].

When electromigration occurs in interconnections, the electromagnetic emission from the integrated circuit increases as shown in Figure 1, with significant increase in the near field electromagnetic field, and the functionality of other parts of an integrated circuit will be affected, and this could render malfunction of the IoT system.

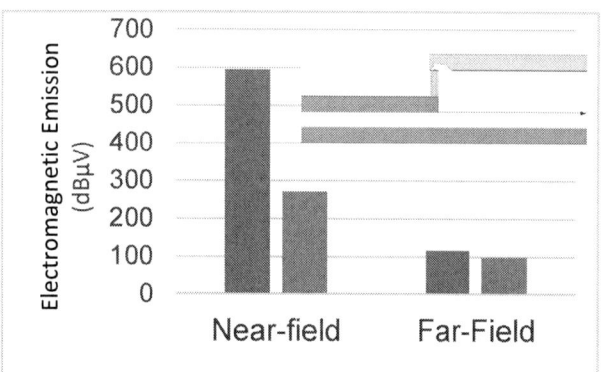

Figure 1. Electromigration and electromagnetic emission from an interconnect in VLSI. Blue and orange colors represent the interconnect with and without electromigration damage respectively. The insert shows the interconnect structure under study

Conclusion

In short, there are many reliability issues that need to be addressed with the inevitable coming of IoT. As IoT must be cost effective for the general consumers, low cost and yet high reliability of sub-system and the entire IoT system become a necessity and the current reliability engineering methodology will be challenged. More research works are needed in the field of reliability for IoT.

.

References

[1] ZdNet. [Online]. Available: http://www.zdnet.com/article/iot-devices-will-outnumber-the-worlds-population-this-year-for-the-first-time/

[2] "Internet of Things – Global Technological and Societal Trend", Ed. Ovidiu Vermesan and Peter Friess, River Publisher Series in Communications, 2011

[3] Tan Cher Ming and Hou Yuejin. "Changing Reliability Physics of Interconnect from Micro- to Nanotechnology", in Ceramic Integration and Joining Technologies: From Macro to Nanoscale, John Wiley, 2010

[4] Eishi Ibe, Hitoshi Taniguchi, Yasuo Yahagi, Ken-ichi, Shimbo, and Tadanobu Toba, "Scaling Effects on Neutron-Induced Soft Error in SRAMs Down to 22nm Process", 2009 Workshop on Dependable and Secure Nanocomputing

[5] E. Petersen, "Single event analysis and prediction," in Proc. IEEE NSREC Short Course, 1997.

[6] J.F. Ziegler, "Terrestrial cosmic rays", IBM J. Res. Devel. 40, 1996, 19-39

[7] C. M. Tan, N. Raghavan, and A. Roy. "Application of Gamma Distribution in Electromigration for Submicron Interconnects," Journal of Applied Physics, vol. 102, no. 10, pp. 103703, 2007.

[8] U. Narula, C.M. Tan, C.S. Lai, "Growth Mechanism for Low Temperature PVD Graphene Synthesis on Copper Using Amorphous Carbon," Scientific Reports, vol. 7, p. 44112, 2017. DOI: 10.1038/srep44112

[9] C. M. Tan, W. Li, Z. Gan, and Y. Hou. Applications of finite element methods for reliability studies on ULSI Interconnections. UK: Springer Verlag, 2011

Vertical Nanowire Tunneling Field-Effect Transistors adopting Core-shell Structure with Strain Effects

Jun-Sik Yoon, *Member, IEEE,* and Chang-Ki Baek, *Member, IEEE*

Abstract— Tunneling field-effect transistor (TFET) is one of the promising candidates to substitute conventional MOSFET for ultra-low power applications. TFETs obey band-to-band tunneling, thus attaining sub-60-mV/dec of the subthreshold swing. However, there is a trade-off of small on-state currents even at high operation voltage. In this work, the SiGe core-shell nanowire TFETs are introduced by showing superior DC characteristics compared to other silicon-based TFETs.

I. INTRODUCTION

Conventional MOSFETs, currently feasible under 10-nm technology node by attaining ultra-sharp fins [1], are facing several bottlenecks such as trade-off between performance and power consumption [2], and self-heating effects [3]. Especially, the trade-off arises from the scaling limitation of subthreshold swing (*SS*) below 60 mV/dec.

Tunneling field-effect transistors (TFETs), on the other hand, follow band-to-band tunneling (BTBT) which enables sub-60-mV/dec of *SS* for several orders of drain currents (I_{ds}), suitable for low power applications [4]. However, the conventional TFETs still have small on-state currents (I_{on}). They require ultra-sharp junction profile and well-aligned gate and junction to increase the DC performance [4], [5]. In this work, using 3-D numerical simulation, the SiGe core-shell nanowire TFETs are proposed.

II. DEVICE STRUCTURE AND SIMULATION METHOD

The concept of core-shell structure has been introduced through electron-hole bilayer [6], bottom-up process [7], and nanotube structure [8]. But both the electron-hole bilayer and the nanotube structure require four terminals and complex device design, which increase the parasitic capacitances at the metal-lines. The core-shell structure using bottom-up process suffers from the ambipolar effect because the distance between the core and the drain is close.

Meanwhile, the proposed core-shell structure in this work consists of the core source regions surrounded by the shell channel regions with a certain intrinsic layer at the top (Fig. 1). Different from the conventional p-i-n (or n-i-p) TFETs, the core-shell TFETs have the tunneling direction parallel to the gate electric field. In addition, the vertical nanowire structure can increase the tunneling area due to its great device density [9]. Feasibility of the vertical nanowire core-shell structure has been explained in the previous work [10].

This research was supported by the Ministry of Science and ICT, Korea, under the ICT Consilience Creative Program (IITP-2017-R0346-16-1007) supervised by the Institute for Information & communications Technology Promotion and also supported by IC Design Education Center.

J.-S. Yoon and C.-K. Baek are with the Department of Creative IT Engineering and Future IT Innovation Laboratory, Pohang University of Science and Technology, Pohang 37673, Korea (e-mail: junsikyoon@postech.ac.kr, baekck@postech.ac.kr).

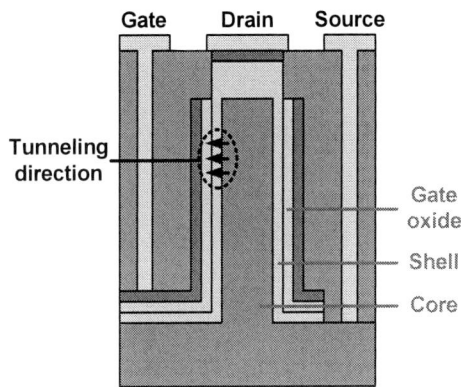

Figure 1. Schematic diagram of the core-shell TFET. Low-k dielectric layer and metal-lines are also specified.

All the core-shell TFETs were analyzed using Sentaurus [11]. Drift-diffusion transport coupled with Poisson and carrier continuity equations were calculated self-consistently. Kane's nonlocal BTBT model [12], Fermi-Dirac distribution, and bandgap narrowing model were used. Carrier mobility was calculated using Masetti and Lombardi models. Shockley-Read-Hall and Auger recombination models were also adopted. Deformation potentials for Si and SiGe were calculated to consider the strain effects of the energy bands. All the simulated core-shell TFETs have the doping concentration of 10^{20}, 10^{19}, and 10^{15} cm^{-3} for source, drain, and channel regions, respectively, which were uniform and abrupt.

III. RESULTS AND DISCUSSION

The core-shell TFETs have better DC performance than do the conventional TFETs (Fig. 2a). The conventional TFETs have the same tunneling area irrespective of the nanowire length (H_{NW}) because the carrier tunneling happens only at the source/channel interface. On the other hand, the core-shell TFETs have the BTBT at the nanowire sidewalls, which can increase the I_{on} greatly as the H_{NW} increases.

Not only does the tunneling area, but also the tunneling electric field is larger for the core-shell TFETs (Fig. 2b). The BTB generation rate increases exponentially as the electric field increases according to the Kane model [12]. Because the core-shell structure has the gate electric field parallel to the electric field at the source/channel interface, the core-shell TFETs could attain greater I_{on} along with wider tunneling area.

In addition, the core-shell TFETs can improve the I_{on} much through the greater H_{NW}, smaller equivalent oxide thickness (EOT), and the strain effects by using different material compositions between core and shell regions (Fig. 3). Smaller EOT improves the gate-to-channel controllability, which enables much band-bending at the source/channel interface.

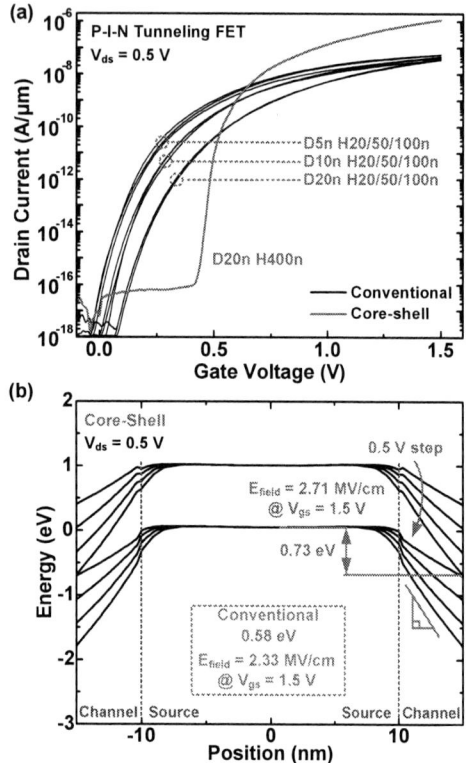

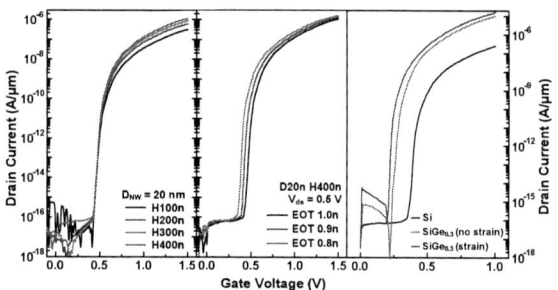

Figure 2. (a) Transfer characteristics of core-shell (red) and conventional (black) TFETs and (b) energy band diagrams of the core-shell TFETs at different gate voltage

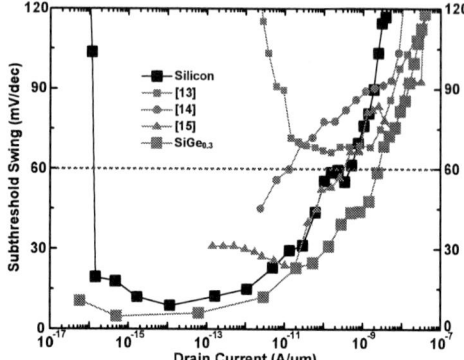

Figure 4. Point *SS* with respect to the drain currents of the TFETs

heterojunction core-shell TFETs are promising for ultra-low power applications.

REFERENCES

[1] H.-J. Cho *et al.*, "Si FinFET based 10nm technology with multi Vt gate stack for low power and high performance applications," *VLSI Symp. Tech. Dig.*, 2016, pp. 1-2.

[2] J.-S. Yoon, C.-K. Baek, and R.-H. Baek, "Process-induced variations of 10-nm node bulk nFinFETs considering middle-of-line parasitics," *IEEE Trans. Electron Devices*, vol. 63, pp. 3399-3405, Sep. 2016.

[3] A. Thean, "Options beyond FinFETs at 5nm node," in *Proc. IEDM Short Course (Nat. Univ. Singapore)*, 2016.

[4] A. M. Ionescu and H. Riel, "Tunnel field-effect transistors as energy-efficient electronic switches," *Nature Review*, vol. 479, pp. 329-337, Nov. 2011.

[5] A. C. Seabaugh and Q. Zhang, "Low-voltage tunnel transistors for beyond CMOS logic," *Proceedings of the IEEE*, vol. 98, pp. 2095-2110, Dec. 2010.

[6] L. Lattanzio, L. De Michielis, and A. M. Ionescu, "Complementary germanium electron-hole bilayer tunnel FET for sub-0.5-V operation," *IEEE Electron Dev. Lett.*, vol. 33, pp. 167-169, Feb. 2012.

[7] J. Nah, E.-S. Liu, K. M. Varahramyan, and E. Tutuc, "Ge-SiGe core-shell nanowire tunneling field-effect transistors," *IEEE Trans. Electron Devices*, vol. 57, pp. 1883-1888, Aug. 2010.

[8] H. M. Fahad and M. M. Hussain, "High-performance silicon nanotube tunneling FET for ultralow-power logic applications," *IEEE Trans. Electron Devices*, vol. 60, pp. 1034-1039, Mar. 2013.

[9] J.-S. Yoon, T. Rim, J. Kim, M. Meyyappan, C.-K. Baek, and Y.-H. Jeong, "Vertical gate-all-around junctionless nanowire transistors with asymmetric diameters and underlap lengths," *Appl. Phys. Lett.*, vol. 105, pp. 102105-1-102105-4, 2014.

[10] J.-S. Yoon, K. Kim, and C.-K. Baek, "Core-shell homojunction silicon vertical nanowire tunneling field-effect transistors," *Nat. Sci. Rep.*, vol. 7, pp. 41142-1-41142-9, Jan. 2017.

[11] *Synopsys, Sentaurus Device User Guide* (Mountain View, CA, 2016).

[12] E. O. Kane, "Theory of tunneling," *J. Appl. Phys.*, vol. 32, pp. 83-91, Jun. 1961.

[13] J.-S. Yoon, K. Kim, M. Meyyappan, and C.-K. Baek, "Bandgap engineering and strain effects of core-shell tunneling field-effect transistors," *IEEE Trans. Electron Devices*, vol. 65, pp. 277-281, Jan. 2018.

[14] Z. X. Chen *et al.*, "Demonstration of tunneling FETs based on highly scalable vertical silicon nanowires," *IEEE Electron Dev. Lett.*, vol. 33, pp. 754-756, Jul. 2009.

[15] H.-Y. Chang, B. Adams, P.-Y. Chien, J. Li, and J. C. S. Woo, "Improved subthreshold and output characteristics of source-pocket Si tunnel FET by the application of laser annealing," *IEEE Trans. Electron Devices*, vol. 60, pp. 92-96, Jan. 2013.

[16] S. H. Kim, S. Agarwal, Z. A. Jacobson, P. Matheu, C. Hu, and T.-J. K. Liu, "Tunnel field effect transistor with raised Germanium source," *IEEE Electron Dev. Lett.*, vol. 31, no. 10, pp. 1107-1109, Oct. 2010.

Figure 3. Transfer characteristics of the core-shell TFETs with different H_{NW} (left), EOT (middle), and SiGe$_{0.3}$ core regions whether or not considering the strain effects.

The SiGe$_{0.3}$ core (Si shell) regions have compressive (tensile) as well as shear strains. These strains decrease the energy bandgaps of the core and the shell regions, thus increasing the BTB generation rates and I_{on} [13].

DC performances among the core-shell and other proposed Si-based TFETs are compared in terms of point *SS* with respect to the I_{ds} (Fig. 4). Point *SS* was calculated by the difference of two adjacent gate voltages over I_{ds}. Overall, the core-shell TFETs have the widest I_{ds} range of sub-60-mV/dec.

IV. CONCLUSION

The core-shell TFETs have their superior DC performance by increasing tunneling area and tunneling electric field compared to the conventional TFETs. In addition, the strain effects induced by the SiGe$_{0.3}$ core and the Si shell modulate the energy bands and increase the I_{on}. Thus, the SiGe/Si

Development of malodor monitoring system based on electronic nose technology

Theerapat Pobkrut[1], Satetha Siyang[2], Treenet Thepudom[2] and Teerakiat Kerdcharoen[1, *]

Abstract— **Odor can affect human health both directly and indirectly specifically malodor. In this work, the portable electronic nose is developed for environmental experiment and it can clearly classify characteristic of each odor source in the factory. Using a zero-grade air as a reference gas and reforming mechanisms of sample testing electronic nose can improve accuracy and efficacy including reduce time of experiment for environmental application. Therefore, this electronic nose can be used for finding the source of malodor that disturb the surrounding villager. The environmental electronic nose can apply to be the internet of things (IoT) device for remotely monitor the odor around the factory and in the village in the future.**

I. Introduction

Electronic nose, also known as E-nose, is wildly used in many fields with related to the smell such as food, beverage, healthcare, agriculture, cosmetic industry, and etc. E-nose consists of the array of gas sensors which can detect wildly volatile in the air or odorous samples. The interaction between gas sensors and target volatiles is comparable to the interaction of human olfactory cells and these reactions cause the change of electrical property, resistance, of the gas sensors. The processing unit of E-nose system detects the electrical property changes of each gas sensor and analyze these data to classify and recognize the kind or sort of that volatile [1]. These processes are the same in human mechanism to recognize the smell [2]. Generally, the principal component analysis (PCA) is used to analyze the odor data of E-nose to classify characteristic of each smell [3]. Among of various materials of gas sensors, metal oxide semiconductor (MOS) is the most reliable and popularity because of their high responsibility and reversible property. However, the most important problem of MOS gas sensor is their high-power consumption to heat the sensing material that has an operating temperature around $250 - 500$ °C [4].

*Research supported by Center of Nanoscience and Nanotechnology.
Teerakiat Kerdcharoen Author is with the Department of Physics and Center of Nanoscience and Nanotechnology, Faculty of Science, Mahidol University, Bangkok 10400, Thailand. (corresponding author to provide phone: +6683 2454499; fax: +662201 5843; e-mail: teerakiat@yahoo.com).
Theerapat Pobkrut Author, is with Department of Physics, Mahidol University, Bangkok 10400, Thailand. (e-mail: T.pobkrut@gmail.com).
Satetha Siyang, and Treenet ThepudomAuthor are with Materials Science and Engineering Programme, Faculty of Science, Mahidol University, Bangkok 10400, Thailand (e-mail: S.satetha@gmail.com and T.thepudom@gmail.com)

Using E-nose in environmental applications is an interesting topic because capability of discriminating and recognizing between variety kind of smells and odors. There are such applications in environment field that E-nose can use such as analyze odor data that cause from abnormal process control or odor control systems malfunction and also use for environmental air quality. Recently, Internet of Things or IoTs play an important role in the development of industrial technology to improve the efficacy of the machines [5]. The combination of Internet of Things, wireless sensor network and electronic nose will enable the factories to monitor the air quality around the factory and increase the quality of life of people nearby.

II. Current Results

Zero-grade air is used to be a reference gas in the environmental application of electronic nose technology. Because it can reduce noise from surrounding air such as contaminant air, humidity, temperature, wind, etc. which can cause the noise in measurement. The result of improvement is shown in Fig 1. The environmental E-nose experiment is setup at 8 places in the factory including inside and outside factory to measure odor from each odorous source sample and compare the odor from factory and the odor in the villages. Sample air is measured for 45 seconds and 150 seconds is set for reference air. The odorous sample is collected by the vacuum pump through 10 m rubber tube from odorous source to the E-nose. Flow rate of air through E-nose system is controlled by the mass flow controller which is 1 liter/minute. The data are sent to monitoring laptop via USB port to collect and save as a raw data to analyses further.

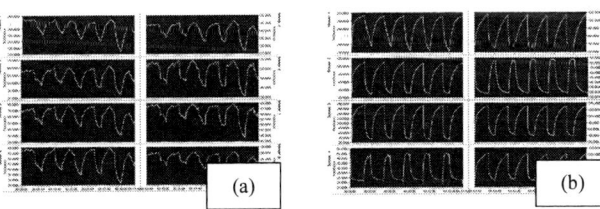

Figure 1. (a) Electronic nose signal without zero grade air be reference signal, (b) Electronic nose signal with zero grade air be reference signal.

TABLE I.   LIST OF SEMICONDUCTOR GAS SENSORS

Sensor	Target gas	Sensor	Target gas
TGS 821	Hydrogen gas	TGS 2600	Air contaminants
TGS 822	Organic solvent vapors	TGS 2602	Air contaminants
TGS-825	Hydrogen sulfide	TGS 2610	LP gas

Sensor	Target gas	Sensor	Target gas
TGS 826	Ammonia	TGS 2620	Solvent vapors

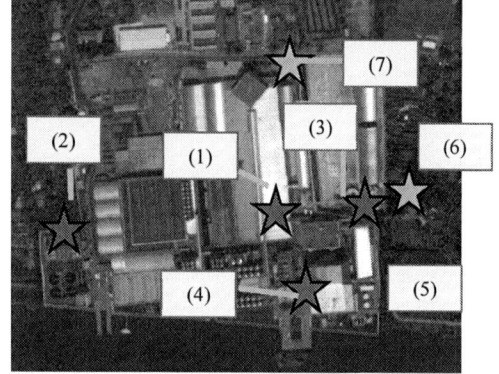

Figure 2. The location of mesurement places in and outside factory which are (1) Sump No.5, (2) Waste water treatment, (3) Stack of the boiler which is using LPG and fuel oil as a fuel, (4) Stack of Multi-wet scrubber of boiler room, (5) Spent grain silo, (6) The village nearby factory, and (7) The market in front of factory.

The first measure is setup in 6 places in the factory which are (1) Sump No.5, (2) Waste water treatment, (3.1) Stack of the boiler which is using LPG as a fuel, (3.2) Stack of Boiler which is using fuel oil as a fuel, (4) Stack of Multi-wet scrubber of boiler room, and (5) Spent grain silo. The result of measurement is shown in Fig.3.

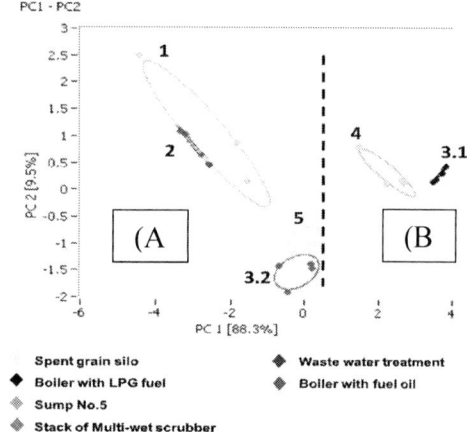

Figure 3. Two-dimensional plot of PCA of 6 odor sources in the factory

Fig. 3 demonstrates that the odor from the factory can separate into two groups which are (A) Strong odor group, and (B) Light odor group which is according to the human test. The PCA result of odor of Sump No.5 and waste water treatment are grouping because these two sources are in the same processes which the water from Sump No.5 flow through the waste water treatment later.

The second experiment is setup in the same 6 places in factory and the addition places outside factory which are (6) The village nearby factory, and (7) The market in front of factory. The result of measurement is shown in Fig. 4.

Fig. 4 demonstrates that the odor from the factory and the odor outside factory can clearly classify. Furthermore, the odor characteristic of (3.2) Boiler with fuel oil and (5) Spent gain silo are grouping because these two places are located nearby.

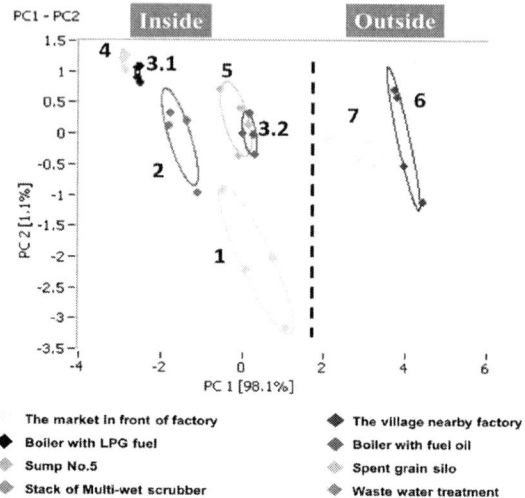

Figure 4. Two-dimensional plot of PCA of 6 odor sources in the factory and 2 places outside factory

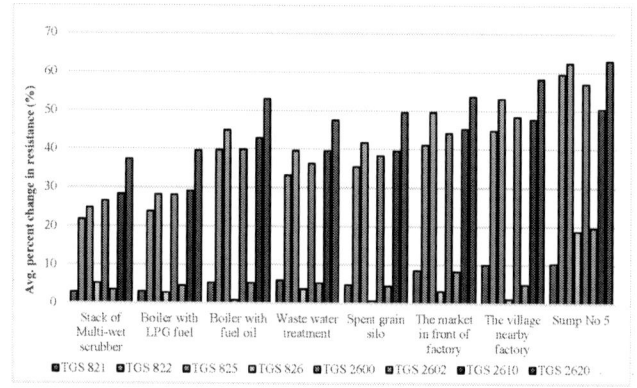

Figure 5. The average percent change in resistance of 8 semiconductor gas sensors which are used in the electronic nose

Fig.5 shows that the average percent change in resistance of the odor from (1) Sump No.5 is the highest among the other sources and the which are similar to human test that the odor from Sump No.5 is the strongest.

REFERENCES

[1] K. Persaud, "Chapter 1 - Electronic Noses and Tongues in the Food Industry," in *Electronic Noses and Tongues in Food Science*, 2016, pp 1-12,

[2] S. Deshmukh, R. Bandyopadhyay, N. Bhattacharyya, R.A. Pandey and A. Jana, "Application of electronic nose for industrial odors and gaseous emissions measurement and monitoring – An overview," in *Talanta*, vol. 144, 2015, pp. 329-340.

[3] C. Wongchoosuk, A. Wisitsoraat, A. Tuantranont and T. Kerdcharoen, "Portable electronic nose based on carbon nanotube-SnO2 gas sensors and its application for detection of methanol contamination in whiskeys," in *Sensors and Actuators B*, vol. 147, 2010, pp. 392-399.

[4] G. F. Fine, L. M. Cavanagh, A. Afonja and R. Binions, "Metal Oxide Semi-Conductor Gas Sensors in Environmental Monitoring," in *Sensors*, vol. 10, 2010, pp. 5469-5502.

[5] J. M. Talavera, L. E. Tobón, J. A. Gómez, M. A. Culman, J. M. Aranda, D. T. Parra, L. A. Quiroz, A. Hoyos and L. E. Garreta, "Review of IoT applications in agro-industrial and environmental fields," in *Computers and Electronics in Agriculture*, vol. 142, 2017, pp. 283-297.

Thermal characterization of polycrystalline diamond using infrared thermal imaging measurement *

Beng Kang Tay, *Member, IEEE*, Qinyu Kong, Laurence Bodelot, Bérengere Lebental, and

Devi Shanker Misra

Abstract— Thermal characterization of polycrystalline diamond based on an infrared thermal imaging technique is proposed. The temperature mapping of polycrystalline diamond under a metal wire heating are recorded using a 15 μm-resolution infrared thermal imaging system. The thermal conductivity of the polycrystalline diamond is derived from the temperature profile using numerical fitting with a 3D heat diffusion model. The thermal conductivity of the polycrystalline diamond is also determined using a 3ω technique. The agreement between the thermal conductivities measured using these two techniques is within 15%.

I. INTRODUCTION

Polycrystalline diamond (PCD) possesses high thermal conductivity and is considered as composing candidate for thermal management material [1]. However, due to the polycrystalline structure of PCD, the localized property of PCD varies at the grains and the grain boundaries. So far, the related studies limit to report the average value of the PCD thermal conductivity. This limitation is the caused by the applied thermal characterization techniques [2, 3]. As the required feature size of heat sink reduces, the study of the localized thermal property becomes more prominent. Therefore, an advanced thermal characterization technique to determine the thermal distribution on polycrystalline diamond is highly required. In this work, the temperature mapping of polycrystalline diamond under metal wire heating is recorded using a high resolution infrared thermal camera. The temperature distribution provides a straightforward way to study the localized heat spreading property of PCD. The averaged thermal conductivity is further derived using numerical fitting with a 3D heat diffusion model. To verify the accuracy of our thermal characterization method, the measured averaged thermal conductivity is compared with the value measured using a 3ω technique.

*Research supported by MOE Tier 1 RG 170/15.

Qinyu Kong and Beng Kang Tay are with NOVITAS, Nanoelectronics Center of Excellence, School of Electrical and Electronic Engineering, Nanyang Technological University, 639798, Singapore (corresponding author to provide phone: +65 67904533; e-mail: EBKTay@ntu.edu.sg).

Laurence Bodelot is with Laboratoire de Mécanique des Solides, Ecole Polytechnique, CNRS, Université Paris-Saclay, 91128 Palaiseau, France.

Bérengere Lebental is with Laboratoire de Physique des Interfaces et des Couches Minces (LPICM), UMR 7647, Ecole Polytechnique-CNRS, Route de Saclay, 91128 Palaiseau, France.

Devi Shanker Misra is with IIa Technologies Pte Ltd. 17, Tukang Innovation Drive, 618300, Singapore.

II. EXPERIMENTAL METHOD

PCD sample is prepared at IIa Technologies Pte Ltd using MPCVD system. The sample surfaces are mechanically polished to achieve a roughness less than 10 nm. A metal wire of 4 mm length and 10 μm width is further deposited on top of the PCD sample to serve as a heater. The PCD sample is suspended by an aluminum holder. A DC current is applied to the metal wire and provides a heating generation rate of 2.27 W. The temperature map of the sample surface is characterized using a FLIR X8400sc Infrared thermal camera. The spatial resolution is 15μm and the minimum detectable temperature change is 25 mK..

III. RESULTS AND DISCUSSION

Figure 1 shows the surface temperature mapping of the PCD sample with a heating power of 2.27 W. The temperature distribution on the diamond surface is not uniform but with lots of dark particles. The temperature profiles along a segment perpendicular to the center of the wire heater (marked by the blue line in the Fig. 2(a)) is further extracted and shown in Fig. 2(b). As the distance from the metal wire heater increases, the temperature of diamond decreases. However, due to the non-uniform defect concentrations within PCD, the temperature profile shows an oscillation with distance duration of around 200 μm.

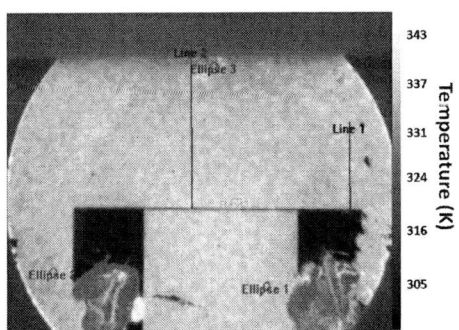

Figure 1. Temperature mapping of the PCD with a heating power of 2.27W

To extract the thermal conductivity of the PCD from the temperature profile shown in Fig. 2(b), we perform numerical fitting using COMSOL Multiphysics. The element size is set as 15μm, which is the same as the spatial resolution of the infrared thermal camera. Figure 3 show the fitting curve of the temperature profile for the PCD sample. The average thermal conductivity of the PCD sample is obtained to be 1010 W/mK. The average thermal conductivity of the PCD sample is further characterized using a 3ω technique. The obtained average thermal conductivity using 3ω technique is 1200 W/mK. The agreement between the thermal conductivity results characterized by two techniques is within 15%.

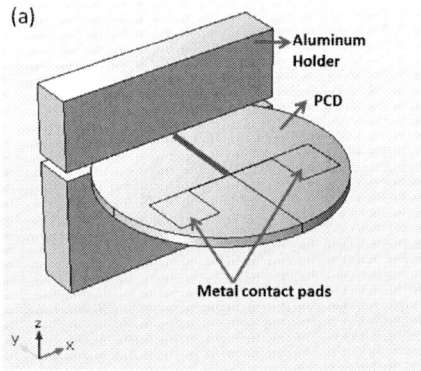

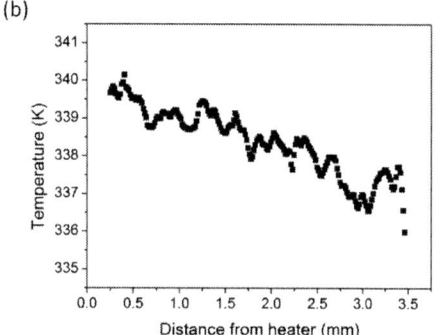

Figure 2. (a) 3D thermal model: the blue line is the segment perpendicular to the middle of the metal wire heater; and (b) The temperature profile along the blue line.

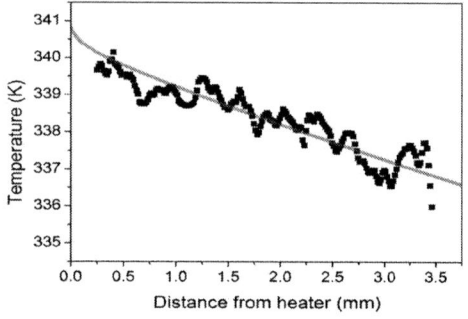

Figure 3. Fitting curve of the temperature profile of PCD.

IV. CONCLUSION

In this study, an infrared thermal imaging method has been proposed for the thermal characterization of polycrystalline diamond. The thermal camera records the temperature profile of polycrystalline diamond under heating by a metal heat wire, and the thermal conductivity is extracted by numerical fitting. The thermal conductivity characterized by the thermal imaging method is comparable with that characterized by the 3ω technique (within 15%). The obtained outcomes from this study provide insightful knowledge on the thermal imaging technique, which gives promising potential to its application for material thermal characterization.

ACKNOWLEDGMENT

This work was supported by MOE Tier 1 RG 170/15 and MOE 2014-T2-2-105. We thank Dr. Ibos for performing the emissivity measurements from CERTES laboratory at Université Paris-EstCréteil.

REFERENCES

[1] Twitchen, D., et al., *Thermal conductivity measurements on CVD diamond.* Diamond and Related Materials, 2001. **10**(3): p. 731-735.

[2] Graebner, J., et al., *Thermal conductivity and the microstructure of state-of-the-art chemical-vapor-deposited (CVD) diamond.* Diamond and Related materials, 1993. **2**(5-7): p. 1059-1063. H. Poor, *An Introduction to Signal Detection and Estimation.* New York: Springer-Verlag, 1985, ch. 4.

[3] Simon, R.B., et al., *Effect of grain size of polycrystalline diamond on its heat spreading properties.* Applied Physics Express, 2016. **9**(6): p. 061302.

A 2DOF MEMS Vibrational Energy Harvester

K. Tao, *Member, IEEE*, L.H. Tang, J. Wu and J.M. Miao, *Member, IEEE*

Abstract— In this paper, a novel two-degree-of-freedom (2DOF) MEMS electromagnetic vibration energy harvesting system is proposed. The dual-mass resonant structure that comprises of a primary mass and an accessory mass is structured on silicon-on-insulator (SOI) wafer by double-sided deep reactive-ion etching (DRIE). Unlike previous 2DOF harvesters, the induction coil is only patterned on the primary mass for energy conversion. By carefully adjusting the weight of accessory mass, the first two resonances of the primary mass can be tuned close to each other while maintain comparable magnitudes. Therefore, both resonances could contribute to energy harvesting that make the system more efficient and adaptive in frequency-variant vibrational circumstances. With the current prototype, two close resonances with a frequency ratio of only 1.19 and comparable peaks are achieved, providing good validation for the modeling results.

I. INTRODUCTION

Energy harvesting technologies are supposed to solve the power supply problem of wireless sensor networks. It has the ability of scavenging and transforming the ambient energy to the electrical power. Regardless of the different working principles, a fundamental issue in vibration-based energy harvesting system is that most of the systems are designed with a linear single-degree-of-freedom system, where maximum output can only be achieved with the narrow bandwidth near their sole resonance [1-3].

One possible method is to obtain broadband energy harvesting by exploiting multiple vibration modes of a multi-degree-of-freedom system [4-5]. Generally, the purpose for designing a broadband multimodal energy harvester is to achieve close resonances with effective magnitudes. However, most of previous studies can only realize resonances quite apart from each other with the second peak considerably less than the first one. The contribution of the high vibration modes is much less than the fundamental mode. On the other hand, a typical electromagnetic energy harvester usually utilizes a permanent magnet as moving proof mass manually assembled to a flexible spring structure. However, it inevitably leads to difficulties in micro-manipulation and weight control of the bulk magnet mass [6-7]. These shortcomings leads to implementation of the 2DOF MEMS electromagnetic harvesters very challenging.

This research is supported by National Natural Science Foundation of China Grant No. 51705429 and the Fundamental Research Funds for the Central Universities. (*Corresponding author: K. Tao*)

K. Tao is with the Ministry of Education Key Laboratory of Micro and Nano Systems for Aerospace, Northwestern Polytechnical University, Xi'an, 710072, China. (e-mail: taokai@nwpu.edu.cn;)

L.H. Tang is with Department of Mechanical Engineering, University of Auckland, 20 Symonds Street, Auckland 1010, New Zealand. (e-mail: l.tang@auckland.ac.nz)

J. Wu and J.M. Miao are with School of Mechanical and Aerospace Engineering, Nanyang Technological University, 50 Nanyang Avenue, 639798, Singapore (e-mail: jwu6@e.ntu.edu.sg; mjmmiao@ntu.edu.sg)

II. DEVICE CONCEPT

In this paper, an architecture with a stationary bulk magnet assembled with moving micro-machined mass and coil is proposed. The bulk magnet is bonded on a fixed suspension plate. The coil is patterned on a 2DOF structure through wafer-level batch-fabrication process with SOI wafers. Thus, the dynamic behavior can be easily adjusted and controlled by modulating the thickness, width and diameter of circular beam and seismic mass. By precisely adjusting the secondary mass, the first two resonances of primary mass can be tuned close to each other with comparable magnitudes. The schematic of 2DOF electromagnetic energy harvester is shown in Figure 1.

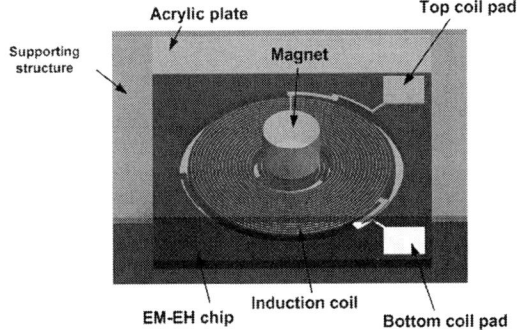

Figure 1. Schematic of the proposed electromagnetic vibration energy harvester with 2DOF MEMS chip

The lumped parameter model of the proposed 2DOF energy harvester is shown in Figure 2. It mainly consists of two subsystems: the primary subsystem (m_1, k_1) and the accessory subsystem (m_2, k_2). The energy conversion damping (η_e) is coupled in the primary subsystem for energy conversion.

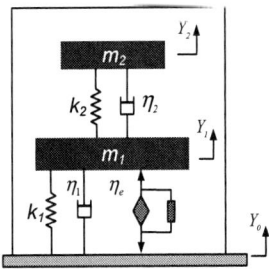

Figure 2. The lumped parameter model of the proposed 2DOF energy harvester

Figure 3 shows the 3D contour of the normalized displacement at different excitation frequencies and mass ratios when the resonant frequencies of the two subsystems is identical. Through parametric analysis, it is found that with a slight weight increase to the original 1DOF system, two close and effective peaks can be achieved. More specifically, when the ratio of the accessory mass to the primary mass is very

978-1-5386-4251-1/18 $31.00 © 2018 IEEE

small and the resonant frequencies of the two subsystems are close, two close and effectives can be obtained. Therefore, the ratio of the accessory mass to the primary mass is set around 0.04. The resonant frequencies are modeled and controlled the same in the current study.

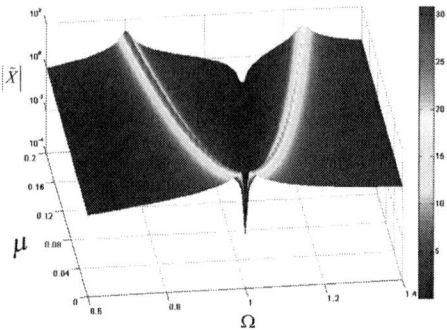

Figure 3. 3D contour of normalized displacement versus excitation frequency and mass ratio when the resonant frequencies of the two subsystems is identical.

III. FABRICATION AND TESTING

Figure 4 shows the fabrication process of the spring-mass structure with coils on a SOI wafer substrate. The SOI wafer has 30 μm thick device layer and 400 μm thick handling layer with 1 μm thick buried oxide layer in the middle. The process includes the Cr/Au coil lift-off process, SiO_2 via etching, frontside feature patterning, frontside and backside DRIE etching and final release process. Figure 5 shows the optical image of a fabricated 2DOF MEMS electromagnetic vibration energy harvester chip in comparison with 10 cents coin. The energy harvesting chip size is 14.5mm×14.5mm×430μm.

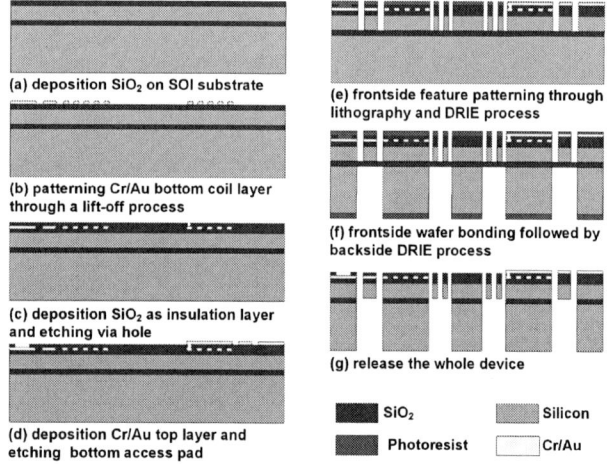

Figure 4. Fabrication process flow of the 2DOF resonant spring-mass structure on a SOI wafer

The fabricated chip is then attached onto an acrylate holder for wire bonding and testing. A swept frequency from 300 to 420 Hz was applied with a load resistance of 10 MΩ. Figure 6 shows the time domain response when a swept frequency is applied. It can be seen that two close peaks of 326.5 Hz and 391 Hz has been obtained. The voltage amplitude is found to be 3.9 and 6.5 mV, respectively. The testing results have a good agreement with the previous modeling prediction.

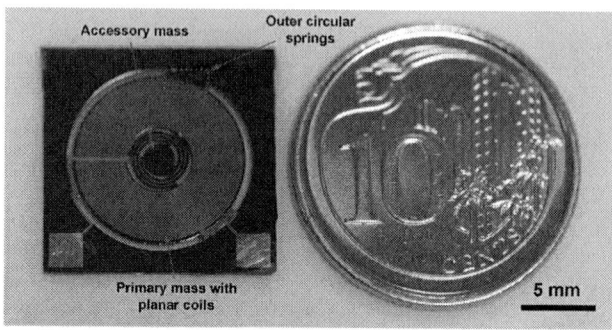

Figure 5. Fabricated 2DOF MEMS electromagnetic vibration energy harvester chip in comparison with 10 cents coin

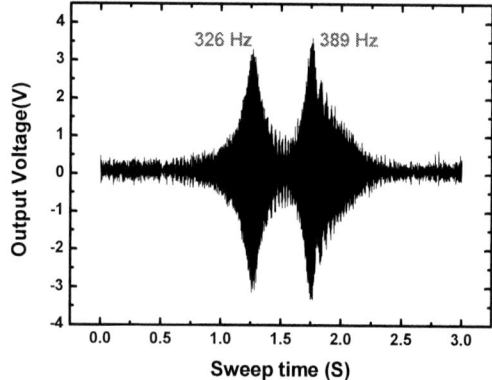

Figure 6. Time domain response of the MEMS electromagnetic vibration energy harvester chip with a swept frequency from 300 to 420 Hz

In summary, this work presents the design, fabrication and characterization of a novel 2DOF MEMS electromagnetic vibration energy harvesting system. The device is constructed on a SOI wafer by standard MEMS fabrication process. From the experimental results, two close and comparable peaks are achieved with the fabricated device. This work provides a possible method of implementation of a MEMS-based 2DOF multimodal energy harvesting device.

REFERENCES

[1] L.H. Tang, Y. Yang, and C.K. Soh, Broadband Vibration Energy Harvesting Techniques, *Advances in Energy Harvesting Methods*, Springer, pp. 17-61, 2013.

[2] K. Tao, J. Wu, L. Tang, L. Hu, SW Lye and J. Miao, "Enhanced electrostatic vibrational energy harvesting using integrated opposite-charged electrets", J. Micromech. Microeng. Vol. 27, p. 044002, 2017.

[3] S. Zhou, L. Zuo. Nonlinear dynamic analysis of asymmetric tristable energy harvesters for enhanced energy harvesting. *Communications in Nonlinear Science and Numerical Simulation*, Vol. 61, pp. 271-284, 2018.

[4] K. Tao, L.H. Tang, J. Wu, S.W. Lye, H.L. Chang and J.M. Miao, "Investigation of Multimodal Electret-based MEMS Energy Harvester with Impact-induced Nonlinearity", J. Microelectromech. Syst. pp. 1-13, 2018, DOI: 10.1109/JMEMS.2018.2792686.

[5] K. Tao, J. Wu, A.G.P. Kottapalli, D. Chen., Z. Yang., G. Ding., S. W. Lye and J.M. Miao, "Micro-patterning of resin-bonded NdFeB magnet for a fully integrated electromagnetic actuator", Solid State Electronics, 138, 66-72, 2017

[6] D.P. Arnold, Review of microscale magnetic power generation, IEEE Trans. Magn. 43(2007) 3940-3951

[7] K. Tao, J. Wu, L. Tang, X. Xia, S.W. Lye, J.M. Miao and X. Hu, "A novel two degree of freedom MEMS electromagnetic vibration energy harvester", J. Micromech. Microeng. Vol. 26, p. 035020, 2016.

HCI and NBTI Reliability Simulation for 45nm CMOS using Eldo

Jaafar, A., N. Soin, S. Wan Muhammad Hatta

Abstract— **Technology scaling on CMOS transistor causes the reliability issues is the main concern by most researchers. HCI and NBTI are the main effects that degrade the CMOS transistor. An analysis of different stages for ring oscillator is analyzed based on HCI and NBTI effect. The acceptable supply voltage for CMOS transistor also analyzed for designer to set the limitation of operational voltage for their design. Eldo simulation shows that the HCI and NBTI degradation on CMOS are stable for 11 and 23 stages of ring oscillator.**

I. INTRODUCTION

Ring oscillator is a digital temperature sensor that widely used to sense the degradation on transistor level up to Fiel Programmable Gate Array (FPGA) level. Ring oscillator is widely used because of low-cost of design and does not required external circuit to construct. Basically ring oscillator is made for odd number of inverters to generate an oscillation with an enable input or reset input to start the oscillation cycles as low as three stages [1] ring oscillator. The performance degradation [2] can be observed through the frequency degradation of the ring oscillator using reliability simulator.

The objective of this paper is to propose the best stages for ring oscillator on 45nm technology node. This paper also focused on acceptable supply voltage range for stable CMOS condition.

The paper is organized as follows. Section II and Section III describes the HCI and NBTI degradations effects respectively. Section IV introduces the simulation process and results obtain using Eldo simulator. The results and analysis is presented in Section V. Finally Section VI presents the conclusions and future work recommendation.

II. HCI

CMOS size are scaling down from time to time, causes the channel of electric field increased apparently, thus enhance the Hot Carrier Injection (HCI) effect. During high electric field, the carrier obtained sufficient energy to become hot carrier. The hot carrier generates interface state for Si-SiO2 interface [3], or new oxide traps, or charge traps in gate oxide. All of these occurrence lead to performance degradation of NMOS transistor.

III. NBTI

Negative Bias Temperature Instability (NBTI) degradation effect refers to stress condition at high temperature on PMOS transistor imposing negative gate voltage. The PMOS transistor generates new interface traps at Si-SiO2 interface [4], thus electric potential will expand at the interface. After that the fixed charge on the interface states causes the positive holes capture which leads the negative shift of threshold voltage [5]. A Reaction-Diffusion (R-D) model is well explained by Wenping Wang about principle of NBTI effect [2].

IV. SIMULATION ON ELDO

A reliability simulation can be simulated using Eldo simulator. Eldo simulator is very useful to predicit the lifetime degradation due to HCI and NBTI effects. CMOS model library of 45nm is used to run the fresh simulation obtained from Predictive Technology Model (PTM) [6] for fresh and aged devices with different stages of ring oscillator and different supply voltage, Vdd. The simulation output is recorded at different temperature range which represents ambient temperature of the devices exposed. At low temperature, the degradation effect discovered are HCI while at high temperature the degradation effect is NBTI. In addition the stability of the stages also discovered when variation of number of stages for ring oscillator and different supply voltage. The simulation process discovered up to 10 years aging in order to determine the sustainability of the CMOS level at 45nm technology node.

V. RESULT AND ANALYSIS

45nm CMOS ring oscillators are simulated using Eldo for different number of stages with different temperature exposed. The degradation effect simulated involved HCI and NBTI mechanisms. Fig 1 shows the frequency degradation for 3 stages ring oscillator using 0°C, 20°C, 27°C, 40°C, 60°C, 80°C and 100°C using Vdd=1V. It shows that the trend of frequency degradations is not stable for temperature equal to 20°C and 27°C. This shows that the 3 stage [7] ring oscillator is not suitable to run on room temperature.

Jaafar, A. is with Faculty of Electronic and Computer Engineering, Universiti Teknikal Malaysia Melaka, Malaysia. (corresponding author to provide phone: 606-555-2000; fax: 606-331-6247; e-mail: anuarjaafar@utem.edu.my).

N. Soin and S. Wan Muhammad Hatta are with Department of Electrical Engineering, University of Malaya, Malaysia. (e-mail: norhayatisoin@um.edu.my; sh_fatmadiana@um.edu.my).

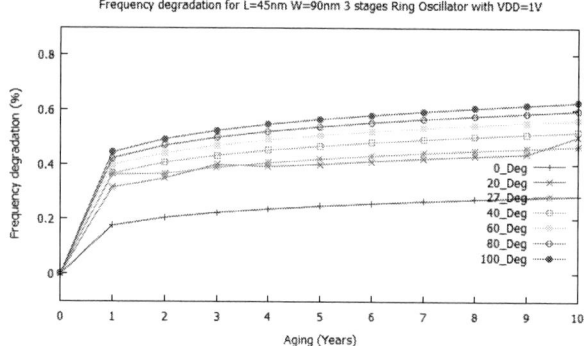

Figure. 1. Frequency degradation for 3 stages ring oscillator.

Meanwhile for 11 and 23 stages ring oscillator produce a reliable frequency degradation trend as shown in Fig 2 and Fig 3. As stated by R. Christoph et al. the most stable stages for FPGA is 23 stages. This shows that for CMOS level, 11 stages is promising the frequency degradation due to HCI and NBTI but it cannot compromise that it will be the most stable stage for FPGA.

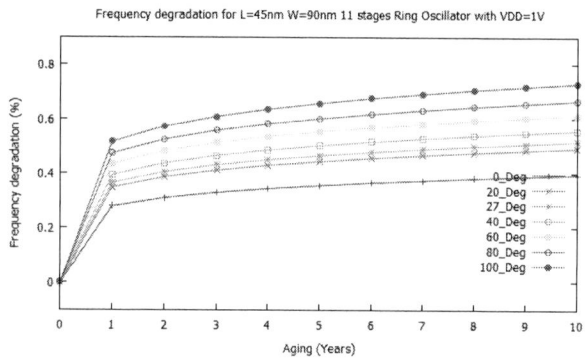

Figure. 2. Frequency degradation for 11 stages ring oscillator.

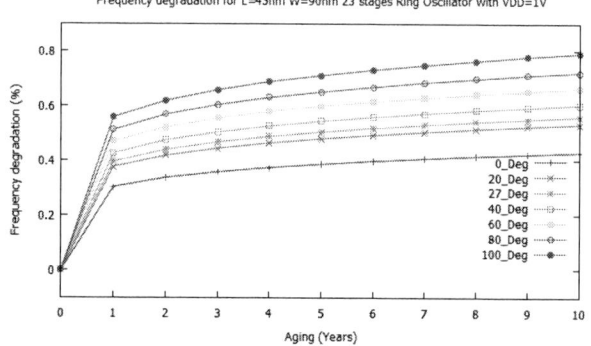

Figure. 3. Frequency degradation for 23 stages of ring oscillator.

Fig 4 shows the frequency degradation for 23 stages ring oscillator with different Vdd for 27°C. It shows that the Vdd values are acceptable for 1.0V, 1.5V, 2.0V and 2.5V. However when Vdd is equal to 3.0V, the frequency degradation of ring oscillator is contradict for lower Vdd value. This shows the Vdd value must not more than or equal to 3.0V.

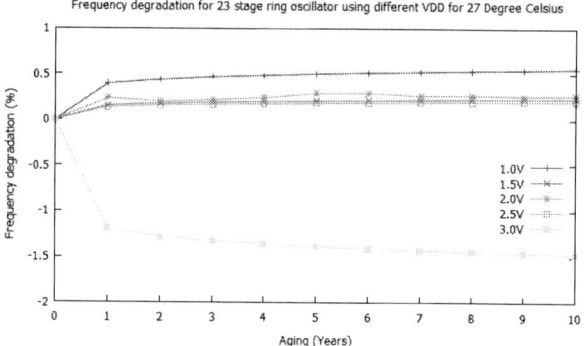

Figure. 4. Frequency degradation for 23 stages ring oscillator for different supply voltages at 27°C.

VI. CONCLUSION

It is recommended that further research be undertaken in the following areas: Further experimental investigations on FPGA are needed to validate and verify the different between fresh and aged device for HCI and NBTI degradation effect. The degradation effect on experiment will provide related findings for different stages of ring oscillator and the implication of current and power consumption itself.

ACKNOWLEDGMENT

The authors would like to acknowledge the faculty at University of Malaya and Faculty of Electronic and Computer Engineering, Universiti Teknikal Malaysia Melaka (UTeM) 76100 Melaka Malaysia for the contributions and reviewers that provide positive feedback to complete this paper. Grant is Postgraduate Research Fund (PG025-2015A).

REFERENCES

[1] T. T. H. Kim, P. F. Lu, K. A. Jenkins, and C. H. Kim, "*A Ring-Oscillator-Based Reliability Monitor for Isolated Measurement of NBTI and PBTI in High-k/Metal Gate Technology*," IEEE Trans. Very Large Scale Integr. Syst., vol. 23, no. 7, pp. 1360–1364, 2015.

[2] H. Zhang, C. Liu, T. Wang, H. Zhang, and C. Zeng, "*Analysis of hot carrier and NBTI induced device degradation on CMOS ring oscillator*," 2013 3rd Int. Conf. Consum. Electron. Commun. Networks, CECNet 2013 - Proc., pp. 141–144, 2013.

[3] M. Cho, P. Roussel, B. Kaczer, R. Degraeve, J. Franco, M. Aoulaiche, T. Chiarella, T. Kauerauf, N. Horiguchi, and G. Groeseneken, "*Channel hot carrier degradation mechanism in long/short channel n-FinFETs*," IEEE Trans. Electron Devices, vol. 60, no. 12, pp. 4002–4007, 2013.

[4] Y. Cao;, Y. Yang, W. He, X. Zheng, X. Ma, and Y. Hao, "*Recovery of PMOSFET NBTI at different cycles*," 2014 12th IEEE Int. Conf. Solid-State Integr. Circuit Technol., pp. 1–3, 2014.

[5] J. Y. Jia, F. Xue, P. Liu, J. Tien, A. Cai, F. Dhaoui, P. Singaraju, F. Hawley, and J. McCollum, "*NBTI life time of a high voltage PMOS FET*," in Proceedings of the International Symposium on the Physical and Failure Analysis of Integrated Circuits, IPFA, 2013, pp. 678–681.

[6] Y. K. Cao, "*Predictive Technology Model*," 2017. [Online]. Available: http://ptm.asu.edu/. [Accessed: 30-Jul-2017].

[7] M. Naouss and F. Marc, "*On-line sensing for healthier FPGA systesm*," FPGA '10 Proc. 18th Annu. ACM/SIGDA Int. Symp. F. Program. gate arrays, pp. 239–248, 2010.

A Compensation Method for Long-term Zero Bias Drift of MEMS Gyroscope Based on Improved CEEMD and ELM

H. Y. Gu, X. X. Liu, B. L. Zhao, and H. Zhou

Abstract— In order to eliminating the long-term zero bias drift of MEMS gyroscope efficiently, a multi-scale processing method is proposed by utilizing signal decomposition. At first, an improved complete ensemble empirical mode decomposition (Improved CEEMD) is used to decompose the original signal into a series of stationary modes; then the distinct sub-series are clustered based on the sample entropy, and extreme learning machine (ELM) based model is used to train the sub-series; finally, the desired results can be obtained after de-noise and compensation. To verify the method, MEMS gyroscope CRG20 has been chosen for an hour test, and the experiment shows that zero bias drift reduced from $0.0706°/s$ to $0.0076°/s(1\text{-}\sigma)$ within temperature range of -40 ℃ to 70℃.

Keywords-MEMS gyroscopes drift; Improved complete ensemble empirical mode decomposition; Extreme learning machine

I. INTRODUCTION

MEMS gyroscopes are widely used in various fields such as navigation, guidance and other measurement applications owing to low power consumption, low cost and the potential for miniature dimensions. However, the precision and long term performance can be easily affected by fabrication imperfections and temperature[1]. In recent years, many literatures focus on signal filtering and drift compensation for eliminating the zero bias drift and random errors of gyroscope. Xiyuan Chen et al reported a thermal induced drift model of fiber optic gyroscope using bounded EEMD and extreme learning machine, and reduced bias instability from $0.26°/s$ to $0.026°/s$(Allan variance) [2]. Bingbo Cui et al reported an improved hybrid filter for fiber optic gyroscope signal de-noising based on EMD and forward liner prediction, and reduced the noises by about 75% [3]. However, many of them only research the temperature drift modeling or linear filtering, and it needs to be extended.

In this paper, a method for drift modeling and signal filtering of MEMS gyroscope based on improved CEEMD and extreme learning machine is proposed. The algorithms are described in Section II; the modeling and filtering procedure

are shown in Section III; Section IV shows the experiment verification; at last the conclusion is given in Section V.

II. ALGORITHMS

A. Improved CEEMD

Empirical mode decomposition (EMD) is an adaptive method to analyze non-stationary signals stemming from nonlinear systems. It can decompose the original signal into a series of functions called intrinsic mode functions (IMFs). Because of the original algorithms exist mode mixing, residual noise and spurious mode problems, the improved algorithm for CEEMD is adopted. Let $E_k(\cdot)$ be the operator which obtains the *k-th* mode of EMD , $M(\cdot)$ be the local mean operator , $w^{(i)}$ be a zero mean unit variance Gaussian white noise and the constant β_k can be set as $\beta_k = \varepsilon_k std(r_k)$[4]. The algorithms can be described as follows: (1) generate and calculate the local means of I realizations: $x^{(i)} = x + \beta_0 E_1(w^{(i)})$ (2) decompose the local means of I realizations by EMD to obtain first residue $r_1 = \langle M(x^{(i)}) \rangle$;(3) Calculate the first mode at $k=1$: $d_1 = x\text{-}r_1$;(4) Estimate the second residue as the average of local means of $r_1 + \beta_1 E_2(w^{(i)})$ and obtain the second mode: $d_2 = r_1\text{-}r_2 = r_1\text{-}\langle M(r_1 + \beta_1 E_2(w^{(i)})) \rangle$;(5) For $k=3,...,K$, calculate the *k-th* residue $\langle M(r_{k\text{-}1} + \beta_{k\text{-}1} E_k(w^{(i)})) \rangle$;(6) Compute the *k-th* mode $d_k = r_{k\text{-}1}\text{-}r_k$. Finally, the series of $d_1, d_2, ..., d_k$ the final results of decomposition. The flowchart of this algorithms is illustrated in Fig 1.

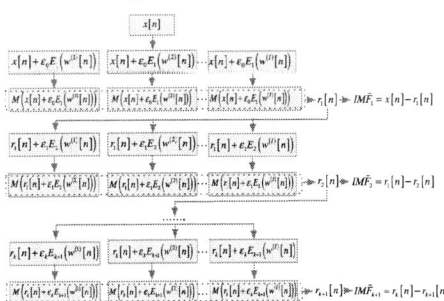

Figure 1. Flowchart describing the improved version of CEEMDAN.

B. Extreme Learning Machine

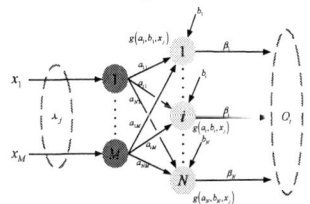

Figure 2. The structure of ELM

H. Y. Gu is with the Institute of Electronic Engineering, China Academy of Engineering Physics, Mianyang, China (phone: 86-138-8199-4139; e-mail: 503287481@qq.com).

X. X. Liu is with the Institute of Electronic Engineering, China Academy of Engineering Physics, Mianyang, China (phone: 86-180-3095-6575; e-mail: liuxx19901015@163.com).

B. L. Zhao is with the Institute of Electronic Engineering, China Academy of Engineering Physics, Mianyang, China (phone: 86-152-8091-6796; e-mail: 18799538@qq.com).

H. Zhou is with the Institute of Electronic Engineering, China Academy of Engineering Physics, Mianyang, China (phone: 86-138-9046-4814; e-mail: chinamems@163.com).

978-1-5386-4251-1/18 $31.00 © 2018 IEEE

Extreme learning machine is a newly developed fast learning method, and which is used for regression with multi-input and single output in this paper. Consider a data set $\mathbf{D} = \left\{ \left(x_j, y_j \right), x_j \in \mathbb{R}^r, y_j \in \mathbb{R}, j = 1, ..., M \right\}$ with M distinct samples.

A standard ELM with $N \leq M$ hidden nodes and activation function $g(x)$ will be applied, which can be expressed as :

$$f_N(x_t) = \sum_{i=1}^N \beta_i \, g(a_i, b_i, x_t) = O_t, t = 1, ..., M \qquad (1)$$

Where , $a_i = [a_{i1}, a_{i2}, ..., a_{iM}]$ is the input weight vector; b_i is the bias of the i-th hidden node; O_t is the predicted output. According to Fig.2, the ELM can be written as (2)-(4)

$$\square\square\square\square\square\square\square\square\square\square \, H\beta = y \qquad (2)$$

$$H = \begin{bmatrix} g(a_1, b_1, x_1) & \cdots & g(a_N, b_N, x_1) \\ \vdots & \cdots & \vdots \\ g(a_1, b_1, x_M) & \cdots & g(a_N, b_N, x_M) \end{bmatrix}_{M \times N} \qquad (3)$$

$$\beta = [\beta_1, ..., \beta_N]^T \qquad (4)$$

$$\hat{\beta} = H^\dagger y \qquad (5)$$

$$\beta = (H^T H)^{-1} H^T y \qquad (6)$$

Where, β is the output weight vector, y is the output vector, H is the hidden layer output matrix, $H^\dagger$ is the Moore-Penrose generalized inverse of H. Due to the input weights and biases are generate randomly, the learning of ELM is to solve β by using the least square method as (5), and β can be presented as (6) .

III. SIGNAL CONTROL SYSTEM

A flowchart of the signal processing procedure is shown in Fig.3.

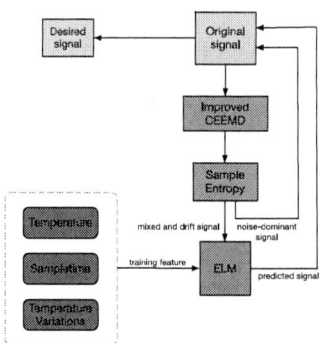

Figure 3. Flowchart of signal processing procedure

Firstly, the original signal is decomposed into IMFs by CEEMD, then we can get the sample entropy of each IMFs, and cluster the IMFs in three signal group as noise-dominate signal, drift signal and mixed signal (mixed with noise and drift); then the noise-dominant signal is filtered out of original signal directly, and the mixed and drift signal is modeling by two independent ELM with three training features to avoid interference; finally, after the original signal is compensated by predicted signal, the desired signal is obtained.

IV. EXPERIMENTAL VERFICATION

To verify the proposed signal processing method, CRG20 has been chosen to test in the temperature range from -40℃ to

70℃ in 1 hour, and the 10000 points test data is collected to PC through the evaluation board which can test six gyroscopes simultaneously. The CRG20 and evaluation board is shown in Fig.4.

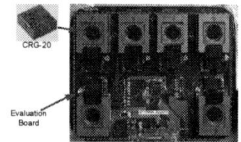

Figure 4. The evaluation borad and testing CRG-20

Fig.5 is the decomposition results of original signal. The original signal is decomposed into 12 IMFs, and clustered the IMFs into three group based on their sample entropy.

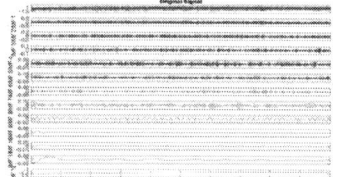

Figure 5. IMFs decomposed by CEEMD

According to the flowchart which is shown in Fig.3, the de-noise and compensation result is shown in Fig.6. The final result shows that the zero bias drift in 1hour reduced from $0.0706°/s$ to $0.0076°/s(1 - \sigma)$.

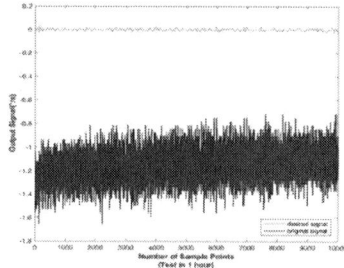

Figure 6. Final results of the processing

V. CONCLUSION

In this paper, a method for compensate the long-term zero bias drift of MEMS gyroscope based on improved CEEMD and extreme learning machine is proposed. This method can reduce the drift from $0.0706°/s$ to $0.0076°/s(1 - \sigma)$, which can improve the long term performance of MEMS gyroscopes effectively.

REFERENCES

[1] Perlmutter M, Robin L, "High-performance, low cost inertial MEMS: A market in motion!"[C]//Position Location and Navigation Symposium (PLANS), 2012 IEEE/ION. IEEE, 2012: 225-229.

[2] Xiyuan Chen, Bingbo Cui, "Efficient modeling of fiber optic gyroscope drift using improved EEMD and extreme learning machine,"[J]. Signal Processing 128(2016) 1-7.

[3] Cui Bingbo, Xiyuan Chen, "Improved hybrid filter for fiber optic gyroscope signal denoising based on EMD and forward linear prediction,"[J]. Sensors and Actuators A 230(2015) 150-155.

[4] Marcelo A. Colominas, "Improved complete ensemble EMD: A suitable tool for biomedical signal processing," Biomedical Signal Processing and Control[J].14 (2014) 19-29.

Evolution of the Physics and Stochastics of Failure in Ultra-Thin Dielectrics – From SiO₂ to Advanced High-K Gate Stacks

Nagarajan Raghavan[*], *Member, IEEE*

Abstract— **Dielectric breakdown in logic devices has been a subject of intense study for several decades. With changing dielectric thickness due to downscaling of complementary metal oxide semiconductor (CMOS) technology as well as the shift from SiO₂ to other high permittivity dielectric materials, there is a noteworthy change in the physics of failure and the statistical trend of soft, progressive and hard breakdown in oxide films. This study presents a brief summary comparing the physical mechanisms of breakdown and associated stochastics of the failure time distribution in SiO₂ and high-κ (HfO₂). It is clearly evident that there is a continuous need for more research into oxide breakdown, given the shift towards 2D layered dielectrics such as hexagonal boron nitride in the near future, with markedly different breakdown dynamics.**

I. INTRODUCTION

Of the various failure mechanisms that are being studied in logic devices, time dependent dielectric breakdown (TDDB) is one of the critical mechanisms. It refers to the time based "random" generation of defects that eventually form a percolation path shorting the gate and the channel region of the transistor. Typically, the time to breakdown distribution in thick SiO₂ films follows the Weibull distribution very closely given the weakest link nature of breakdown [1] – [4]. In recent times, we notice an increasing deviation from the standard Weibull trend in several advanced metal gate – high-κ stacks [5] – [7]. The reasons behind these deviations have to be probed in detail. In the section that follows, the various aspects of breakdown where differences exist in SiO₂ and high-κ dielectrics will be discussed in brief from a physical viewpoint.

II. COMPARING BREAKDOWN PHYSICS IN SiO₂ WITH HIGH-K

A. Role of Oxide Microstructure (Grain Boundaries)

Silicon oxide and oxynitride are both perfectly amorphous materials, in comparison to hafnium oxide, which generally tends to be polycrystalline in nature. The presence of grain boundaries (denoted as GB) (defect concentrated regions) makes it easier for oxygen vacancy defects in the vicinity to migrate towards the boundary, as a result of which breakdown events tend to be located more at GB sites [8, 9].

B. Role of Gate Electrode

With the shift from SiO₂ to high-κ for effective oxide thickness scaling, there was a corresponding change in the gate electrode from poly-silicon to TiN/TaN/NiSi as well. This resulted in several extrinsic failure issues in the dielectric such as migration of the metal atoms / ions forming conductive filament during hard breakdown of the stack [10, 11] In contrast, the mechanism of hard breakdown in SiO₂ / SiON was more of an epitaxial Si protrusion into the oxide

[*]This research supported by the SUTD Start-Up Research Grant SREP15108. N. Raghavan is with the Engineering Product Development (EPD) Pillar, Singapore University of Technology and Design (SUTD), 8, Somapah Road, Singapore - 487372

(Corresponding author: +65 6499 8756; E-mail: nagarajan@sutd.edu.sg).

leading to effective oxide thinning as well as fully oxygen depleted Si core inside the SiO₂ [12, 13]. The metal migration mechanism effectively reduces the reliability margin for high-κ stacks as there is no progressive degradation of the dielectric after the TDDB event (as observed for SiO₂) [10]. Instead, we observe a catastrophic hard breakdown event due to quick migration and punch through of the oxide by the metal atoms / ions.

C. Dual Layer Dielectric in High-κ Stacks

With hafnium oxide being deposited and not intrinsically grown on the silicon substrate, there is almost always a trace amount of sub-stoichiometric SiO$_x$ that grows between the high-κ (HK) and the Si substrate. As such, the oxide is a dual-layer material stack with HfO₂ and SiO$_x$ (interfacial layer (IL)). With different activation energies for defect generation (4.4eV for HfO₂ and 2.3-2.7eV for SiO₂ [14]) and different field distribution in the HK and IL, breakdown events tend to be preferentially nucleate within the IL layer [14]. Even after the breakdown of the IL layer (soft BD), the field increment in the HK layer is not sufficient to cause a complete breakdown of the dual layer stack. As a result, at the circuit level, the occurrence of a hard breakdown (HBD) event is very rare and in general, circuit failure is only caused by multiple soft breakdown (SBD) events within the IL [14].

D. Defect Clustering in the High-κ

Recent atomistic studies have shown that in HfO₂, the presence of a vacancy defect reduces the activation barrier for another defect to be created in its vicinity and this reduction is quite significant [15, 16]. As a result, it is being speculated that defect nucleation may not be a Poisson random process in the HK anymore. Instead, defect generation is spatially localized and evolves like a "dendrite" from one defect to a complete chain of defects. This is very different from random defect generation in SiO₂, where no clustering phenomenon has been observed so far. The existence of defect clustering has an impact on the percolation model and it results in the breakage of the traditional Weibull slope – oxide thickness linear relationship [17].

III. COMPARING TDDB STATISTICS IN SiO₂ WITH HIGH-K

A. Invalidity of the Weibull Distribution

With defect clustering in the HK and the existence of a dual layer oxide stack, the area scaling laws and the standard Weibull distribution no longer apply to TDDB in HK stacks [18]. As a result, the clustering model has been recently proposed by Wu *et al.* [19] to better describe the statistics of oxide breakdown in the HK. It has one additional parameter (α) when compared to the standard Weibull distribution. Here, α denotes the cluster factor which empirically quantifies the extent to which defects nucleate and evolve in a clustered fashion.

B. Extrapolation of Accelerated Test Results

Often times, we just assume an acceleration model for oxide time to failure, which follows a simple power law as a function of gate voltage stress or ξ-model or the 1/ ξ model (where ξ denotes the electric field) [20]. However, the extrapolation using these empirical laws ceases to be technically true considering the different conduction mechanisms for electron tunneling in the dielectric at operating and accelerated stress conditions [21]. In general, the TDDB tests are performed at "stress" voltages high enough for electrons to be in Fowler-Nordheim (F-N) tunneling regime, while at operating conditions, the mechanism is more of direct tunneling. As such, except for highly defective stacks where trap assisted tunneling might dominate across the entire voltage regime, the differences in the conduction mechanism will imply the need for a shift in the extrapolation model at intermediate stress conditions, more so given that defect generation and breakdown is not just field driven, but also affected by carrier fluence [22, 23].

IV. CONCLUSION

A brief review of the differences in the breakdown physics and statistics of SiO_2 and high-κ dielectrics is provided in this work from a fundamental material point of view. The differences in the statistics and physics of the different dielectric stacks is further compounded by the paradigm shift in the device geometry from planar MOSFETs to FinFETs and possibly, gate all around devices in the future. These geometrical changes affect the spatial field patterns greatly, which further influences the statistics of TDDB. There is a lot of scope for further work to be done in understanding the physical and stochastic nature of TDDB, soft, progressive and hard breakdown in advanced dielectric stacks and this also requires the use of high-end physical failure analysis tools such as the transmission electron microscope (TEM) and the scanning tunneling microscope (STM). Finally, the use of the clustering model, which seems to be the ideal distribution for future TDDB data analysis both from technology point of view (wafer scale variations in oxide thickness) as well as the fundamental physics point of view (correlated vacancy generation mechanisms in HfO_2), has to be phenomenologically justified using analytical derivations as well so that the community is convinced on the need for it.

ACKNOWLEDGMENT

The author would like to acknowledge the support from the SUTD start-up research grant (SREP15108) for funding and logistically supporting this work.

REFERENCES

[1] R. Degraeve, G. Groeseneken, R. Bellens, M. Depas and H.E. Maes, "A consistent model for the thickness dependence of intrinsic breakdown in ultra-thin oxides", *IEEE International Electron Devices Meeting*, pp. 863-866, (1995).

[2] J. Suñé, "New physics-based analytic approach to the thin-oxide breakdown statistics", *IEEE Electron Device Letters*, 22(6), pp.296-298, (2001).

[3] J.H. Stathis, "Percolation models for gate oxide breakdown", *Journal of Applied Physics*, Vol. 86, No. 10, pp.5757-5766, (1999).

[4] E.Y. Wu and R.P. Vollertsen, "On the Weibull shape factor of intrinsic breakdown of dielectric films and its accurate experimental determination. Part I: theory, methodology, experimental techniques", *IEEE Transactions on Electron Devices*, Vol. 49. No. 12, pp.2131-2140, (2002).

[5] E. Cartier, A. Kerber, T. Ando, M.M. Frank, K. Choi, S. Krishnan, B. Linder, K. Zhao, F. Monsieur, J. Stathis and V. Narayanan, "Fundamental aspects of HfO_2-based high-κ metal gate stack reliability and implications on T_{inv}-scaling", *IEEE International Electron Devices Meeting (IEDM)*, 18-4, (2011).

[6] N. Raghavan, K.L. Pey, K. Shubhakar and M. Bosman, "Modified percolation model for polycrystalline high-κ gate stack with grain boundary defects", *IEEE Electron Device Letters*, Vol. 32, No. 1, pp.78-80, (2011).

[7] J. Suñé, S. Tous and E.Y. Wu, "Analytical cell-based model for the breakdown statistics of multilayer insulator stacks", *IEEE Electron Device Letters*, Vol. 30, No. 12, pp.1359-1361, (2009).

[8] K. McKenna, A. Shluger, V. Iglesias, M. Porti, M. Nafría, M. Lanza and G. Bersuker, "Grain boundary mediated leakage current in polycrystalline HfO_2 films", *Microelectronic Engineering*, Vol. 88, No. 7, pp.1272-1275, (2011).

[9] K. Shubhakar, K.L. Pey, N. Raghavan, S.S. Kushvaha, M. Bosman, Z. Wang and S.J. O'Shea, "Study of preferential localized degradation and breakdown of HfO_2/SiO_x dielectric stacks at grain boundary sites of polycrystalline HfO_2 dielectrics", *Microelectronic Engineering*, Vol. 109, pp.364-369, (2013).

[10] K.L. Pey, N. Raghavan, X. Li, W.H. Liu, K. Shubhakar, X. Wu and M. Bosman, "New insight into the TDDB and breakdown reliability of novel high-κ gate dielectric stacks", *IEEE International Reliability Physics Symposium (IRPS)*, pp. 354-363, (2010).

[11] T. Kauerauf, R. Degraeve, M.B. Zahid, M. Cho, B. Kaczer, P.J. Roussel, G. Groeseneken, H.E. Maes and S. De Gendt, "Abrupt breakdown in dielectric/metal gate stacks: A potential reliability limitation?", *IEEE Electron Device Letters*, Vol. 26, No. 10, pp.773-775, (2005).

[12] C.H. Tung, K.L. Pey, L.J. Tang, M.K. Radhakrishnan, W.H. Lin, F. Palumbo and S. Lombardo, "Percolation path and dielectric-breakdown-induced-epitaxy evolution during ultrathin gate dielectric breakdown transient", *Applied Physics Letters*, Vol. 83, No. 11, pp. 2223-2225, (2003).

[13] N. Raghavan, A. Padovani, X. Li, X. Wu, V.L. Lo, M. Bosman. L. Larcher and K.L. Pey, "Resilience of ultra-thin oxynitride films to percolative wear-out and reliability implications for high-κ stacks at low voltage stress", *Journal of Applied Physics*, Vol. 114, Issue 9, 094504, (2013).

[14] N. Raghavan, A. Padovani, X. Wu, K. Shubhakar, M. Bosman, L. Larcher and K.L. Pey, "The "buffering" role of high-κ in post breakdown degradation immunity of advanced dual layer dielectric gate stacks", *IEEE International Reliability Physics Symposium (IRPS)*, 5A-3, (2013).

[15] S.R. Bradley, A.L. Shluger and G. Bersuker, "Electron-injection-assisted generation of oxygen vacancies in monoclinic HfO_2", *Physical Review Applied*, Vol. 4(6), 064008, (2015).

[16] S.R. Bradley, *Computational Modelling of Oxygen Defects and Interfaces in Monoclinic HfO_2* (Doctoral dissertation, UCL (University College London)), (2016).

[17] R. O'Connor, G. Hughes and T. Kauerauf, "Time-dependent dielectric breakdown and stress-induced leakage current characteristics of 0.7-nm-EOT HfO_2 pFETs", *IEEE Transactions on Device and Materials Reliability*, Vol. 11(2), pp. 290-294, (2011).

[18] E.Y. Wu, B. Li, J.H. Stathis and R. Achanta, "Multiple breakdown phenomena and modeling for non-uniform dielectric systems", *IEEE International Electron Devices Meeting (IEDM)*, 34-7, (2014).

[19] E.Y. Wu, B. Li and J.H. Stathis, "Modeling of time-dependent non-uniform dielectric breakdown using a clustering statistical approach", *Applied Physics Letters*, Vol. 103(15), 152907, (2013).

[20] E.Y. Wu and J. Suñé, "On voltage acceleration models of time to breakdown—Part I: Experimental and analysis methodologies", *IEEE Transactions on Electron Devices*, Vol. 56(7), pp.1433-1441, (2009).

[21] P.E. Nicollian, "Errors in Projecting Gate Dielectric Reliability from Fowler–Nordheim Stress to Direct-Tunneling Operation", *IEEE Electron Device Letters*, Vol. 30(11), pp.1185-1187, (2009).

[22] J. Wu, E. Rosenbaum, B. MacDonald, E. Li, J. Tao, B. Tracy and P. Fang, "Anode hole injection versus hydrogen release: The mechanism for gate oxide breakdown", *IEEE International Reliability Physics Symposium*, pp. 27-32, (2000).

[23] M.F. Li, Y.D. He, S.G. Ma, B.J. Cho, K.F. Lo and M.Z. Xu, "Role of hole fluence in gate oxide breakdown", *IEEE Electron Device Letters*, Vol. 20(11), pp.586-588, (1999).

Strategies in Microfluidic Self-Assembled Nanoparticles

Ciprian Iliescu, Guillaume Tresset, Ming Ni and Cătălin Mărculescu

Abstract— The current work presents the using of microfluidic hydrodynamic flow focusing for enabling an accurate control of self-assembling nanoparticles. A mixture of surfactant and DNA dispersed in 35% ethanol is focused between two streams of pure water in a microfluidic channel. As a result, a rapid change of solvent quality takes place in the central stream, and the surfactant-bound DNA molecules undergo a fast coil-globule transition. Using this method nanoparticles having a hydrodynamic diameter of 70nm with a polydispersity index below 0.2 were achieved. A second method relied on the controlled diffusive mixing of surfactant and DNA solutions through a water stream of tunable width. Using this method the smallest nanoparticles achieved were about 30 nm in hydrodynamic diameter, meaning that most of them contained a single DNA molecule.

I. INTRODUCTION

Nanotechnologies are about to induce profound changes in our society and may have a tremendous impact in various fields spanning from information technology to biomedical sciences. However, scientists have yet to devise strategies to harness the matter at the nanoscale, more specifically, to fabricate with a high throughput functional nanosystems responding to precise specifications.[1] Self-assembly processes are viewed as the best way for the fabrication of 3D-structured nanosystems such as those involved in biomedical sciences.[2] Nanosystems designed to deliver active drugs or genes into target cells,[3, 4] or to label specific tissues for bioimaging,[5] are undoubtedly of great importance for the diagnosis and the treatment of diverse pathologies, and will translate into substantial economic benefits. Microtechnology and especially microfluidics has a strong impact in biomedical field approaching problems related transdermal drug delivery,[6-9] cell culture for tissue engineering applications, [10, 11] label free separation of different cell population [12, 13], isolation of circulating tumor cells[14,15]. Nevertheless, microfluidics star to be applied also in self-assembling of nanoparticle, few recent reviews underlining the role that microfluidic can play especially in the size control and homogeneity of the nanoparticle achieved. [16-18]

*Resrach supported by Centre National de la Recherche Scientifique (CNRS, France) through the PICS program (project ref. #6662).
C. Iliescu is with BIGHEART, National University of Singapore, MD6, 14 Medical Drive, #14-01, Singapore 117599 (corresponding author -e-mail: bigci@nus.edu.sg).
G. Tresset is with Laboratoire de Physique des Solides, CNRS, Univ. Paris-Sud, Université Paris-Saclay, 91405 Orsay Cedex, France
M. Ni is with the School of Biological Sciences & Engineering, Yachay Tech University, Hacienda San José s/n, San Miguel de Urcuquí 100105, Ecuador
C. Marculescu is with National Institute for Research and Development in Microtechnologies, IMT-Bucharest, Bucharest 077190, Romania.

II. DNA SURFACTANT NANOPARTICLES

A. Limitations of the Hydrodynamic Flow Focusing

The hydrodynamic flow focusing method described in [19] was tested for the DNA compaction. Calf Thymus DNA was flow through the central microfluidic channel while a surfactant was used for the side stream (as illustrated in Fig. 1). The experimental results show a large polydispersion index (values over 0.5) and values of the hydrodynamic diameter larger than 200nm (making the nanoparticle less suitable for gene therapy applications). The results can be explained through the difference between linearcharge density of CT-DNA and CMC (5.9e/nm and 2.5 e/n respectively).

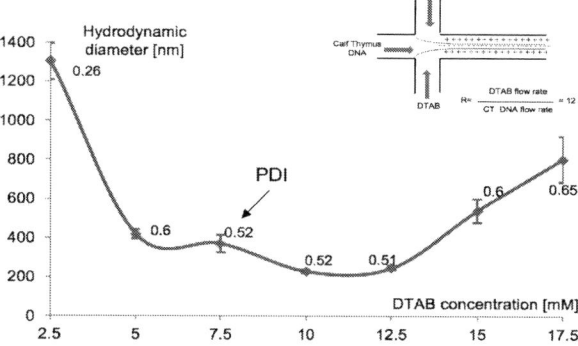

Figure 1. Hydrodynamic flow focusing method applied for Calf Thymus DNA compaction using DTAB

B. Solvent Exchange Using Hydrodynamic Flow Focusing

The method consists of separation of electrostatic interaction (between the cationic headgroup of the surfactant and DNA) from the hydrophobic interaction between the alkyl chains of the surfactant molecules. Practically, the compaction is performed in two steps. First one in bulk, mixing the DNA and surfactant in 35% ethanol solution. The second on chip, focusing the DNA-surfactant-ethanol solution through two water streams, diluting in this way the ethanol concentration and enhancing the hydrophobic interaction between the alkyl chain of the surfactant molecules. In Fig. 3 the variation of the hydrodynamic diameter with the concentration of the surfactant is presented. As a result 60nm DNA-DTAB nanoparticles were achieved with a PDI below 0.2. The method has the advantage of simplicity while the flowing parameters presents a low influence on particle size. Moreover, the presence of ethanol traces in the nanoparticle, used for solubilizing the surfactant molecules, can limit the application of the method for gene therapy.

978-1-5386-4251-1/18 $31.00 © 2018 IEEE

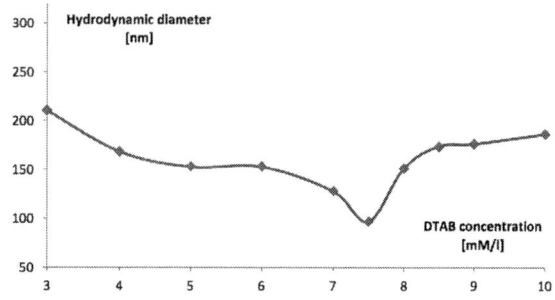

Figure 2. Solvent exchange method: a stream having DNA and DTAB in 35% Ethanol is focused between two water streams.

C. Slow Diffusion Through a Buffer Stream

In order to overcome the disadvantage of the previous method, we design an "organic solvent free method". In this method the adsorption time of the surfactant molecule to the DNA chain was drastically increased by the diffusion of these molecules through a water stream. The layout of the microfluidic device, is depicted in Fig. 3a. The chip presents three inlets for DNA, water and DTAB. The fabrication of the device in glass/silicon technology was preferred, not only due to the robustness [20, 21] of the device, but mainly to the opportunity of reusing the chip after a short cleaning process with NMP. The working principle is relatively simple: a DI water stream is sandwiched between a stream of DNA and another one of DTA B. DNA and DTAB slowly diffused across the water stream. As a result, the surfactant molecules interact one by one in contact with DNA chains. The DNA-DTAB coil-globule transition occurred in a dilute regime, which aloud condensation of only one or few DNA chains in the same nanoparticle. The results presented in Fig. 3b shows that nanoparticles with the hydrodynamic diameter of 30 nm can be achieved. This can be equivalent with the presence of one DNA molecule in one nano-particle.

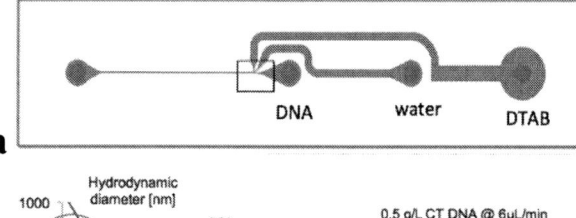

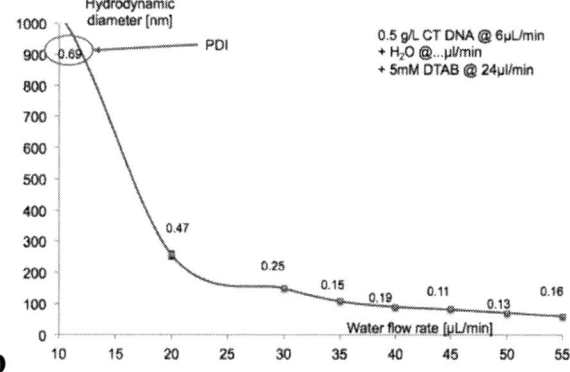

Figure 3. a) Layout of the chip; b) Variation of the hydrodynamic diameter as a function of the water stream flow rate.

III. CONCLUSION

To conclude, we design three methods for self-assembling of DNA- surfactant nanoparticles. A major advantages of the proposed methods are related to the fine control of the hydrodynamic diameter of the nanoparticle using flowing conditions and /or concentration of the solutions and also the control of the uniformity of the particles (in most polydispersion index is below 0.2).

REFERENCES

[1] X.F. Zhao, W.T. Winter, "Cellulose/cellulose-based nanospheres: perspectives and prospective," *Ind. Biotech.*, vol. 11, pp. 34-43, 2015.

[2] G. Tresset, C. Iliescu, "Microfluidics-directed self-assembly of DNA-based nanoparticles," *Inform. MIDEM*, vol. 46, pp.183-189, 2017.

[3] C. Iliescu, G. Tresset, "Microfluidics-driven strategy for size-controlled DNA compaction by slow diffusion through water stream," *Chemistry of Materials*, vol. 27, pp. 8193-8197, 2015.

[4] C. Iliescu, C. Mărculescu, S. Venkataraman, B. Languille, H. Yu, G. Tresset, "On-chip controlled surfactant–DNA coil–globule transition by rapid solvent exchange using hydrodynamic flow focusing," *Langmuir*, 29 (44), pp. 13125-13136, 2014

[5] M. Ni, G. Tresset, C. Iliescu, "Self-assembled polysulfone nanoparticles using microfluidic chip," *Sensor Actuat B: Chem*, vol. 252, pp.458-462, 2017.

[6] F. S. Iliescu, A. P. Sterian, M. Petrescu, "A parallel between transdermal drug delivery and microtechnology," *U. Politeh. Buch. Ser. A*, vol. 75, no. 3, pp. 227-236, 2013.

[7] D. Resnik, et al, "In vivo experimental study of noninvasive insulin microinjection through hollow Si microneedle array," *Micromachines-Basel*. vol. 9(1), p. 40. 2018.

[8] F.S. Iliescu, S. Paunica, D. Vrtacnik, A.R. Bobei, "A double softlithography method for processing of noa63 microneedles arrays," *U. Politeh. Buch. Ser. B*, vol. 79, pp. 121-132, 2017.

[9] H. Zhang, J. K. Jackson, M. Chiao, "Microfabricated drug delivery devices: design, fabrication, and applications," *Adv. Funct. Mat.*, vol. 27(45), 1703606, 2017.

[10] F. Yu, F.S. Iliescu, C. Iliescu, "A comprehensive review onperfusion cell culture systems," *Inform. MIDEM*, vol. 46(4), pp.163-175, 2017.

[11] F. Yu et al., "A perfusion incubator liver chip for 3D cell culture withapplication on chronic hepatotoxicity testing," *Sci. Rep.-UK* vol. 7, p. 14528, 2017.

[12] C. Iliescu, G. Tresset, G. Xu, "Dielectrophoretic field-flow method for separating particle populations in a chip with asymmetric electrodes," *Biomicrofluidics*, vol. 3, (4), p. 044104, 2009.

[13] F. S. Iliescu, et al. "Continuous separation of white blood cell from blood in a microfluidic device," *U. Politeh. Buch. Ser. A*, vol. 71, no. 4, pp. 21-30, 2009.

[14] J. Jiang et al., "An integrated microfluidic device for rapid and high sensitivity analysis of circulating tumor cells," *Sci. Rep.-UK*, vol. 7, 2017.

[15] I. Cima et al., "Label-free isolation of circulating tumor cells in microfluidic devices: Current research and perspectives," *Biomicrofluidics*, vol. 7(1), p. 011810, 2013.

[16] J. Ma, et al. "Controllable synthesis of functional nanoparticles by microfluidic platforms for biomedical applications–a review," *Lab Chip*, vol. 17 (2), pp. 209-226, 2017.

[17] R.-J. Yang, et al, "Micromagnetofluidics in microfluidic systems: A review," *Sensor Actuat. B: Chem*, vol. 224, pp. 1-15, 2016.

[18] M. Lu et al., "Microfluidic hydrodynamic focusing for synthesis of nanomaterials," *Nano Today*, 2016

[19] G. Tresset, et al. , "Fine control over the size of surfactant–polyelectro-lyte nanoparticles by hydrodynamic flow focusing," *Anal. Chem.*, vol. 85 (12), pp. 5850-5856, 2013.

[20] C. Iliescu, J. Miao, F.E.H. Tay, "Optimization of an amorphous silicon mask PECVD process for deep wet etching of Pyrex glass," *Surf. Coat. Tech.*, vol. 192(1), pp.43-47, 2005.

[21] C. Iliescu, et al "A practical guide for the fabrication of microfluidic devices using glass and silicon," Biomicrofluidics, vol. 6(1), p. 016505, 2012.

Electrical analysis of InGaAs-based planar and tri-gate nMOSFET with S/D resistance dependencies at different drain biases

N. A. F. Othman, S. W. M. Hatta, *Member, IEEE* and N. Soin, *Member, IEEE*

Abstract—This paper studies the electrical analysis of InGaAs-based planar and tri-gate nMOSFET and the influence of the source/drain (S/D) resistance, R_{sd} on the current-voltage (I_d-V_g) relation at different drain biases (V_{ds}). It is found that the tri-gate nMOSFET simulated at high V_{ds} has shown better performance compared to planar nMOSFET simulated at low V_{ds}. As the R_{sd} is reduced, the drain current of both planar and tri-gate devices increases. The on-current to off-current (I_{on}/I_{off}) ratio of the devices also increases as the R_{sd} reduced. Tri-gate nMOSFET shows significant improvement as the I_{on}/I_{off} ratio is 10^3 higher than the planar nMOSFET device.

I. INTRODUCTION

Group III-V compound semiconductors are being actively studied for the past decade as they show potential applications in the future technology of complementary metal oxide semiconductors (CMOS) devices to replace the silicon (Si) channel [1-3]. These compounds, e.g. indium gallium arsenide (InGaAs) has high electron mobility and is regarded as one of the promising compound for channel material, especially for n-type CMOS devices. In this paper, the electrical analysis of InGaAs-based planar and tri-gate nMOS field effect transistor (nMOSFET) were studied. The impacts of source/drain (S/D) resistance, R_{sd} at different drain biases were also discussed.

II. METHODOLOGY

A. Device Structure

The devices under study is a 60 nm technology node MOSFETs, where the performance of the n-type device were studied. The MOSFETs were simulated using two dimensional (2D) for planar device, and 3D for tri-gate device, as shown in Fig. 1. Table I gives the list of parameters and values used in the simulation. As compared to earlier work, the half-structure is simulated in order to obtain more accurate results [4].

B. Simulation Details

In this work, 2D and 3D simulation is performed to generate a planar and tri-gate nMOSFET, respectively. The device formation steps includes the initialization setup, epitaxy growth, device formation, and device meshing. The device formation is illustrated in Fig. 2. The devices are simulated in different drain biases (V_{ds}), that is V_{ds} = 0.05 V in linear mode and V_{ds} = 0.5 V in saturation mode. The impacts of R_{sd} on the device electrical characteristics were studied. The R_{sd} were varied from 50 Ω, 100 Ω, and 150 Ω. The figure of merits involved in this work are threshold voltage (V_t), on-current (I_{on}) and off-current (I_{off}). V_t of the devices were extracted from the linear mode, while I_{on} and I_{off} were extracted from the

saturation mode. In this study, the Synopsys Sentaurus TCAD simulator is used, which includes the high-field velocity saturation model and low-field electron mobility of InGaAs for InGaAs-based planar and tri-gate nMOSFET simulation.

TABLE I. VALUES AND PARAMETERS USED IN THE SIMULATION

Parameters	Values	
	Fixed	*Variables*
Gate length, L_{gate}	0.06 μm	-
Channel depth, T_{gate}	0.04 μm	-
Fin width, W_{fin}	0.04 μm	-
Oxide thickness, t_{ox}	2.67 nm	-
Source/drain thickness, T_{sd}	0.03 μm	-
Drain bias, V_{ds}	-	V_{ds} = 0.05V, V_{ds} = 0.5V
Source/drain resistance, R_{sd}	-	50Ω, 100Ω, 150Ω

III. RESULTS AND DISCUSSION

This section discusses the simulation result of InGaAs-based planar and tri-gate nMOSFET. Fig. 3 illustrates the current-voltage (I_d-V_g) characteristics of the planar and tri-gate nMOSFET at different V_{ds}. It is observed that the I_d-V_g characteristics of the planar nMOSFET at higher drain bias shows better performance compared to tri-gate nMOSFET with the same bias. At higher drain, the tri-gate nMOSFET exhibit higher I_{on} compared to low drain biased of planar nMOSFET. At gate voltage (V_g) = 0 V, the current generated for tri-gate nMOSFET is lower compared to planar nMOSFET. This shows that the tri-gate nMOSFET generate lower I_{off}. This enhancement is favourable for tri-gate nMOSFET since the device produce high I_{on} with low I_{off}. Fig. 4 illustrates the cross sections of the fin of the tri-gate nMOSFET, showing the electron current distributions at different drain bias. It is found that the current is highly distributed for higher drain bias compared to lower drain bias. This is expected due to the generation of I_{off} in higher drain bias devices. Moreover, Fig. 5 shows the impacts of source/drain resistance (R_{sd}) on the I_d-V_g characteristics for planar and tri-gate nMOSFET. From the figure, it is observed that as the R_{sd} reduced, the drain current increased particularly at higher V_{gs}. This is due to the current behaviour, by which the current will increased in reduced R_{sd} as discussed in [5-7]. The inset of Fig. 5 presents the V_t with respect to the R_{sd}. It is found that the V_t increased as the R_{sd} is reduced. The V_t for planar devices are seen to have lower values as compared to tri-gate devices. Fig. 6 demonstrated the I_{on}/I_{off} ratio which represents the device power consumptions. It is observed that the device ratio is higher at low R_{sd} for both planar and tri-gate nMOSFET. For tri-gate devices, the ratio is 10^3 higher than the planar device. This shows that the tri-gate devices are better in

---

Research supported in part by the University of Malaya/Ministry of Science, Technology and Innovation Science Fund Grant (UM/MOSTI Science Fund) under Grant UM.0000072/HMT.SF, and in part by the Post-Graduate Research Fund (PPP) under Grant PG329-2016A.

The authors are with the Department of Electrical Engineering, University of Malaya, Kuala Lumpur 50603, Malaysia (phone: +603-7967-5205; e-mail: sh_fatmadiana@um.edu.my).

handling high power applications due to their better power consumptions.

IV. CONCLUSION

A 2D and 3D analysis on the 60 nm nMOSFET was done to determine the impacts of R_{sd} on the performance of InGaAs-based planar and tri-gate device, respectively. It is found that tri-gate nMOSFET exhibits better drive current at high drain bias compared to planar device at low drain bias. Furthermore, the drain current shows improvement as the resistance reduces. The on-current to off-current ratio also shows that the tri-gate device improved by 10^3 magnitude as the resistance reduced compared to planar device, indicating that the tri-gate devices can be used in high power applications due to their good power consumptions.

APPENDIX

Figure 1.   (a) 2D illustration for planar nMOSFET and (b) 3D illustration for tri-gate nMOSFET

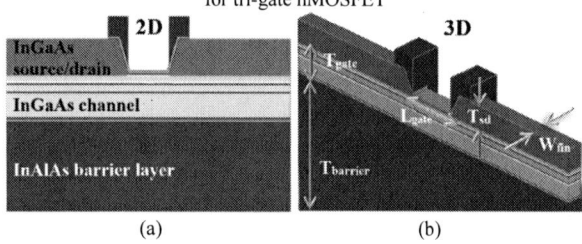

(a)   (b)

Figure 2.   Project flow of the 2D and 3D simulations

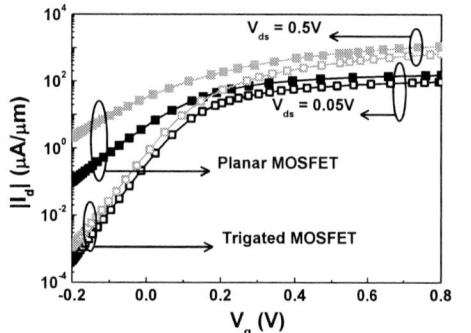

Figure 3. I_d-V_g graph of planar and tri-gate nMOSFET at different V_{ds}.

Figure 4.   Electron current distributions at different V_{ds} for tri-gate nMOSFET

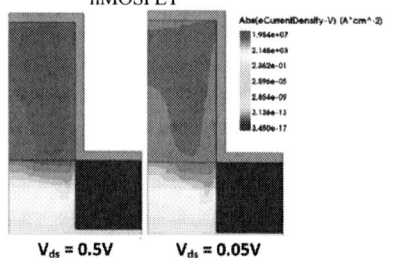

Figure 5. I_d-V_g graph for various R_{sd} at $V_{ds} = 0.05$V

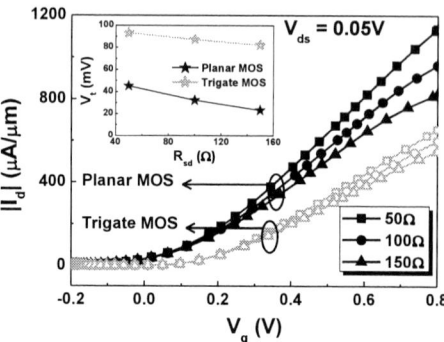

Figure 6. I_{on}/I_{off} ratio of planar and tri-gate nMOSFET for various R_{sd}

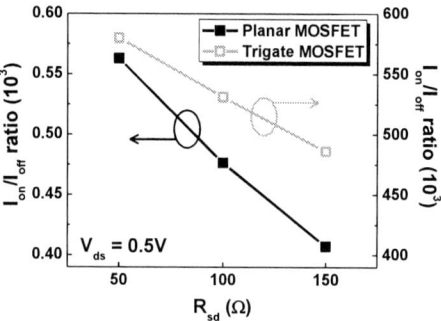

ACKNOWLEDGMENT

The authors acknowledge financial support from the Ministry of Science, Technology and Innovation (MOSTI) for Science Fund Grant Scheme (Science Fund – SF012-2015) and the University of Malaya for Postgraduate Research Fund (PPP – PG329-2016A).

REFERENCES

[1]   J. Ma, J. F. Zhang, Z. Ji, B. Benbakhti, W. D. Zhang, X. F. Zheng, *et al.*, "Characterization of negative-bias temperature instability of Ge MOSFETs with GeO$_2$/Al$_2$O$_3$ Stack," *IEEE Transactions on Electron Devices*, vol. 61, pp. 1307-1315, 2014.

[2]   J. J. Gu, X. Wang, H. Wu, J. Shao, A. T. Neal, M. J. Manfra, *et al.*, "20–80nm channel length InGaAs gate-all-around nanowire MOSFETs with EOT= 1.2 nm and lowest SS= 63mV/dec," in *Electron Devices Meeting (IEDM), 2012 IEEE International*, 2012, pp. 27.6. 1-27.6. 4.

[3]   M. Radosavljevic, G. Dewey, D. Basu, J. Boardman, B. Chu-Kung, J. Fastenau, *et al.*, "Electrostatics improvement in 3-D tri-gate over ultra-thin body planar InGaAs quantum well field effect transistors with high-K gate dielectric and scaled gate-to-drain/gate-to-source separation," in *Electron Devices Meeting (IEDM), 2011 IEEE International*, 2011, pp. 33.1. 1-33.1. 4.

[4]   X. Wang, A. R. Brown, N. Idris, S. Markov, G. Roy, and A. Asenov, "Statistical threshold-voltage variability in scaled decananometer bulk HKMG MOSFETs: A full-scale 3-D simulation scaling study," *IEEE Transactions on Electron Devices*, vol. 58, pp. 2293-2301, 2011.

[5]   F. A. M. Rezali, N. A. F. Othman, M. Mazhar, S. W. M. Hatta, and N. Soin, "Performance and Device Design Based on Geometry and Process Considerations for 14/16-nm Strained FinFETs," *IEEE Transactions on Electron Devices*, vol. 63, pp. 974-981, 2016.

[6]   J. Joseph and R. Patrikar, "Impact of Fin Width and Graded Channel Doping on the Performance of 22nm SOI FinFET," in *VLSI Design and Test*, ed: Springer, 2013, pp. 153-159.

[7]   S. Chabukswar, D. Maji, C. Manoj, K. Anil, V. R. Rao, F. Crupi, *et al.*, "Implications of fin width scaling on variability and reliability of high-k metal gate FinFETs," *Microelectronic Engineering*, vol. 87, pp. 1963-1967, 2010.

Fluxless Flip Chip Interconnects for MEMS Devices for Heterogeneous Integration

Teck Kheng Lee

Flip chip technologies (FC) is an enabling technology for heterogeneous integration of MEMS devices. FC reduces the package dimension, improves performance as well as enable scaling up of MEMS chips through 3-D packaging. This paper provide an overview of FC technologies for MEMS and identify the interconnect challenges for MEMS packaging. The paper shares an innovative fluxless FC bonding technique known as solid liquid interdiffusion by compressive force (SLICF) for MEMS packaging.

The instantaneous SLICF bonding utilizes a mechanical force to break the Sn oxide layer and allows the submerged body to interact with fresh molten solders to form bonds through solid liquid inter-diffusion. This remove the need of flux and ideal for MEMS packaging for heterogeneous integration. The JIV plugged molten solder in via on flex substrate to enable a fluxless FC bonding for MEMS integration. The fine pitch design rule and foldable of flex substrate enable ease of heterogeneous integration. The bonding architecture enable fluxless FC bonding for heterogeneous integration of MEMS devices to reduce packaging cost and provide a high throughput for heterogeneous integration.

I. INTRODUCTION

Microelectromechanical system (MEMS) and emerging nanoelectromechanical system (NEMS), henceforth both referred to as 'MEMS', are transducer systems that sense or control mechanical, optical or chemical quantities, such as pressure, acceleration, light intensity or fluids. The MEMS device interacts with the environment and converts the response as an electrical interface to the outside world. With the growth of internet-of-things, MEMS transducers have taken a centre stage to convert the signals to useful information for interpretation. The International Technology Roadmap for Semiconductors [1] describes the addition of MEMS functionalities to ICs as an important 'More-Than-Moore' technology. As MEMS fabrication is not compatible with CMOS process, heterogeneous integration is a preferred approach as compare to system-on-chip (SoC) solution. Wire bonding is commonly employed for MEMS packaging. The wire bonding may not be the preferable technique for MEMS packaging as it restrict in miniature due to loop height and fan-out of the wires [2-3]. In addition, the longer traces and wire loop will affect the electrical performance of the package. In contrast, flip-chip packaging provides a number of advantages such as: 1) providing MEMS structures with a covering lid, which is the die/chip itself [4]; 2) opportunities to scale up the number of MEMS chips; and 3) integration of MEMS with others microchips (hybrid integration) [5], such as ASIC, microfluidics, and microgenerators, while maintaining the same X-Y form factor of the package through advanced 3-D stacking technology. In additional, it enhances the electrical performance of the system.

II. FLIP CHIP TECHNOLOGIES AND ITS CHALLENGES

FC is the building block to enable heterogeneous integration for MEMS applications. The first level interconnect typical make use of controlled collapsed chip connection with the use of flux. Unfortunately, the process of solder paste and solder spheres cannot be applied directly to MEMS. One of the main reasons is the flux contaminations on the MEMS chip. Flux is needed for solder paste or solder spheres as it clean and prepare the surface for soldering with other metals such as Cu and Au to form alloy joint. For MEMS application, cleaning flux cannot be tolerable as the deflux process will be catastrophic for the mechanical microstructures. For many MEMS structures which contain complex active/moveable mechanical parts, the DI water cleaning is not acceptable due to stiction issues [6] which will reduce or impede the MEMS performance.

Most fluxless bonding techniques, including inert atmosphere soldering, acid-vapor soldering, hydrogen soldering, laser soldering, and plasma- assisted dry soldering (PADS), use some means of controlled atmosphere, such as shielding atmosphere or plasma, to reduce oxidation for better solderability [7-8]. The challenges lay in the handling, storage, and infrastructure limitations that make most fluxless bonding techniques cumbersome in high volume implementation. As such, there is a need for a low cost high productive fluxless FC interconnect technologies for MEMS packaging

III. INNOVATIONS IN FLUXLESS FC BONDINGS AND ITS ARCHITECTURE

The use of flux has been known to inhibit the wide adoption of FC for MEMS packaging. The fluxless solid liquid inter-diffusion by compressive force (SLICF) to form interconnects will be shared in this paper to ease the heterogeneous integration with MEMS.

A. Solid Liquid Interdiffusion by Compressive force

Instantaneous fluxless bonding by solid liquid interdiffusion by compressive force is shown in Fig 1. SLICF bonding works on the basic of using a mechanical force to break the oxide layer and overcome the surface tension and the weight of the insertion body. Once inserted, the submerged body interacts with fresh solder to form bonds via solid-liquid interdiffusion. The bonds are mechanically interlocked upon cooling [9-10]. Figure 1 illustrates the working principle of SLICF bonding. This fluxless bonding technique does not rely on wetting behavior, thus allowing instantaneous soldering. It also eliminates the need for the fluxing and reflow process for fine pitch and MEMS applications.

B. Joint-in-via architecture

Figure 2(a) shows a new substrate architecture known as Joint-in-via (JIV) architecture for fluxless bonding. It also serves as means of closing the technology gap between the die and the substrate by Lee et al. [10-11]. JIV architecture shifts the current copper pad designs to a micro-via architecture using existing substrate technologies. JIV architecture consolidates the landing pads, micro-vias, and flip chip joint into one common element, thereby saving valuable substrate space for high-density routing. The pad pitch resolution is improved by eliminating the need for a solder mask and reducing the inter-pad space using laser ablation. JIV architecture enables a flip chip pad pitch as low as 70μm, with receiving via pads of 50μm on a flex laminate (Figure 3a). This eliminates the need for costly RDL processes without compromising the joint reliability and package geometry and enable heterogeneous integration through folding.

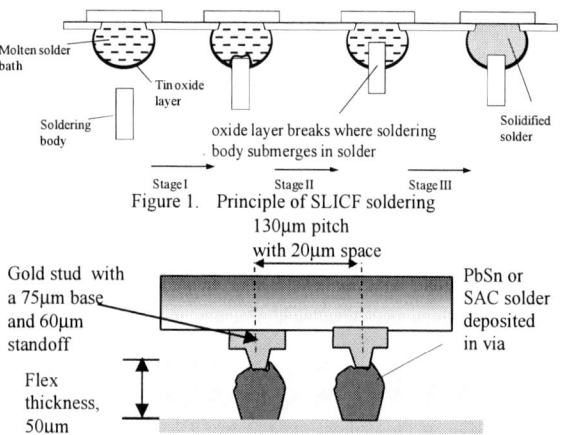

Figure 1.   Principle of SLICF soldering

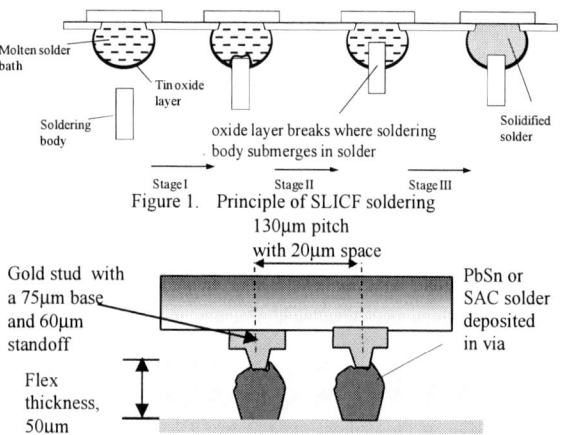

Figure 2.   Schematic illustration of SLICF bonding with the JIV architecture

The solderpaste is plugged in the via and act as molten solder reservoir for SLICF bonding. A series of Au studs are mechanically bumped on the MEMS device using thermosonic wire bonder. The Au studs are then bonded into corresponding JIV architectures containing SAC solders forming instantaneous interconnects without any controlled environment. Figure 3(b) shows the microstructure of the AuSn$_4$ and AuSn$_2$ formation of typical JIV joint of Au with SAC. The interconnects with conventional underfill survive the standard reliability tests as shown in Table 1

IV.   CONCLUSION

One of the challenges for FC for heterogeneous integration of MEMS devices is the use of flux for soldering. The use of flux limit size of reduction and performance as FC cannot be applied for heterogeneous integration with MEMS. The JIV architecture leverage on the instantaneous SLICF bonding for fluxless FC bonding without any controlled environment. The fluxless SLICF bonding utilizes a mechanical force to break the Sn oxide layer and allows the submerged body to interact with fresh molten solders to form bonds through solid liquid inter-diffusion. This remove the need of flux and ideal for MEMS packaging for heterogeneous integration. The JIV architecture enables a fluxless FC bonding for MEMS heterogeneous integration. The interconnects were characterized and survive the standard reliability testing such

as Temperature Cycle Condition B, Moisture Sensitive Test Level 2 and Thermal shock condition D.

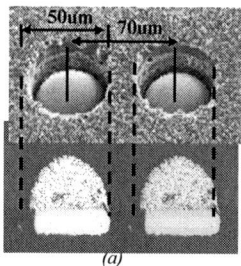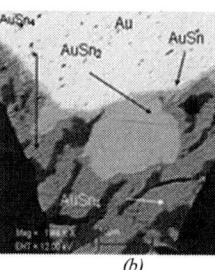

Fig 3(a) Solder plugged into vias of 70um pitch (b) Intefacial microstructures showing AuSn$_4$ and AuSn$_2$ formation of typical JIV joint of Au with SAC

TABLE I.   PACKAGE RELIAIBIILTY BASED ON CONTINUITY RESISTANCE AND C-SAM FR 8 X 9 MM PACKAGE

No	Test Description	Reliability Result
1	MST L2 85°C/60% RH with 3IR reflow 260°C based on J-STD-020	0/15
2	MST L3 30°C/60% RH with 3IR reflow 260°C based on J-STD-020	0/15
3	Temperature cycling condition A –40 to 85°C for 1000 cycles based on EIA/JESD22-A105-B	0/15
4	Temperature cycling condition B –40 to 125°C for 1000 cycles based on EIA/JESD22-A105-B	0/15
5	Thermal Shock Condition D –65 to 150°C for 700 cycles based on JESD22-A106-A	0/15

REFERENCES

[1] ITRS. International Technology Roadmap for Semiconductors, 2013 Edition. The International Technology Roadmap 2013

[2] Chen LT, Cheng WH. A novel plastic package for pressure sensors fabricated using the lithographic dam-ring approach. Sens. Actuators A, Phys. 2009 Jan.149:165–171.

[3] Jemmy Sutanto, Sindhu Anand, Chetan Patel, Jit Muthuswamy. Novel first-level Interconnect Techniques for Flip chip on MEMS Devices. J Microelectromech Syst. 2011 Nov 3; 21(1): 132–144.

[4] Boustedt K, Persson K, Stranneby D. Flip chip as an enabler for MEMS packaging; Proc. 52nd Electron. Compon. Technol. Conf..2002. pp. 124–128.

[5] Basavanhally N, Lopez D, Aksyuk V, Ramsey D, Bower E, Cirelli R, Ferry E, Frahm R, Gates J, Klemens F, Lai W, Low Y, Mansfield W, Pai CS, Papazian R, Pardo F, Sorsch T, Watson P. High-density solder bump interconnect for MEMS hybrid integration. IEEE Trans. Adv Packag. 2007 Nov.30(4):622–628.

[6] Maboudian R, Ashurst WR, Carraro C. Self-assembled mono-layers as anti-stiction coatings for MEMS: Characteristics and recent developments. Sens. Actuators A, Phys. 2000 May 15;82(1-3):219–223.

[7] Miller DC, Zhang W, Bright VM. Microrelay packaging technology using flip-chip assembly; Proc. 13th Annu. Int. Conf. MEMS.2000. pp. 265–270

[8] Singh A, Horsley DA, Cohn MB, Pisano AP, Howe RT. Batch transfer of microstructures using flip-chip solder bump bonding; Proc. Int. Conf. TRANSDUCERS; Chicago, IL. 1997. pp. 265–268.

[9] T.K Lee, Sam Zhang, C.C. Wong, A.C. Tan, "Instantaneous fluxless bonding of Au with Pb-Sn in ambient atmosphere," *Journal of Applied Physics*, 98, 034904.

[10] T.K Lee, Sam Zhang, C.C. Wong, A.C. Tan, "Fluxless Flip Chip Bonding with Joint-in-Via Architecture," *Thin Solid Films*. 504 (2006) pp. 436-440

[11] T. K. Lee, Sam Z, C. C Wong, A. C. Tan, *"Assessment of Fluxless Solid Liquid Interdiffusion Bonding by Compressive Force of Au-PbSn and Au-SAC for Flip Chip Packaging"*, IEEE Trans- CPMT- B, Vol 32, Issue 1 (2009), pp: 116-122.

An Enhanced High-Sensitivity Micro Resonant Thermometer with Axial Strain Amplification Effect

First Q. Shen, Second J. Yang, Third J. B. Xie, Fourth S. Ren, Fifth W. Z. Yuan

Abstract—This paper presents an enhanced high-sensitivity micro resonant temperature sensor with axial stain amplification structure. With the external temperature variation, an amplificated axial strain of micro-resonator vibration beam will be produced because of the materials with different thermal expansion coefficients of the micro resonator and package. This will result in a larger resonant frequency shift of the device with temperature change. The theory and analysis of frequency variation is illustrated with temperature sensitivity achieving 309Hz/K. Experimental test shows that actual frequency variation with temperature change of 20K is about 252Hz/K. Simulation match with measurement moderately and can be utilized to optimally design high-sensitivity temperature sensor before the costly fabrication.

I. INTRODUCTION

Resonant temperature sensors are widely studied for its superiority in sensitivity, digitized signal output and resolution [1-2]. Recently, high-sensitivity resonant temperature sensor based on silicon material has focused on more attention because of its low cost and batch processing. Kose T at al [3] design silicon-on-insulator (SOI)-based resonant temperature sensor bonded on glass layer with the aim of thermal strain amplification to achieve larger frequency shift. Then, through gluing this device to substrate of the package, external temperature variation is detected with sensitivity of about 20Hz/K. Nevertheless, the induced glass middle layer not only delay thermal conduction and increase heat hysteresis from the package layer to the silicon structure layer but leads to stress-amplifying degradation and restrict temperature sensitivity improvement because of its thermal expansion coefficient approaching that of the package layer.

In this paper, we present an enhanced strain-amplifying micro resonator to improve the sensitivity of temperature sensing. The sensor provides an amplified signal output, thanks to the strain-amplifying beam structure directly attached to the package without the middle layer mentioned above. The sensitivity of the temperature sensor can be increased by changing the materials of the package and resonator with larger difference of thermal expansion coefficients.

II. THEORY AND SIMULATION

Fig.1 shows the micro resonator, comprised strain-amplifying structure, resonant beam, electrodes and substrate. Using materials with different thermal expansion coefficients in package and resonator, a larger axial force will be loaded on the fixed ends of the beam, which creates larger frequency shift. Finally, the connection between frequency shift and temperature change is illustrated.

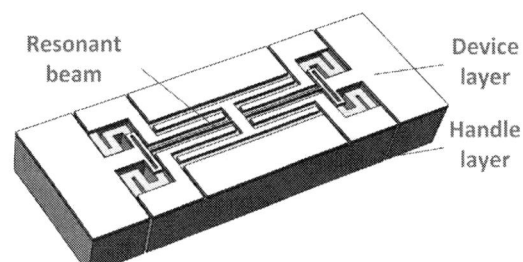

Figure.1 Structure of resonator

This is achieved by aluminum material in package and silicon material in resonator because of a large difference in thermal expansion coefficient of aluminum and silicon ($\alpha_{silicon} = 2.6 \times 10^{-6}$, $\alpha_{aluminum} = 23.1 \times 10^{-6}$).

Fig.2 illustrates the resonant beam structure. The axial load on beam due to temperature change can be described by

$$F = \cfrac{L(\alpha_{Al} - \alpha_{Si})\Delta T}{\cfrac{L_1 + L_2 + L_3}{E_{Al}A_{Al}} + \cfrac{L_1}{E_{Si}W_1h} + \cfrac{L_3}{E_{Si}W_3h} + \cfrac{L_2}{2E_{Si}Wh} + \cfrac{\Delta W}{16E_{Si}W^3h}} \quad (1)$$

$$\Delta W = (W_2 - W_1)^3 + (W_2 - W_3)^3 \quad (2)$$

where E_{Al} and E_{Si} are the Young's modulus of aluminum and silicon, α_{Al} and α_{Si} are the thermal expansion coefficients of aluminum and silicon, h is the thickness of the beam, and ΔT is the temperature change.

As a resonator, the resonant frequency can be described by

$$f = 2\pi\sqrt{m_{eff}/k_{eff}} \quad (3)$$

$$k_{eff} = \frac{EI}{L^3}\int_0^1\left(\frac{d^2\Phi_i}{d\varepsilon^2}\right)^2 d\varepsilon + \frac{F}{L}\int_0^1\left(\frac{d\Phi_i}{d\varepsilon}\right)^2 d\varepsilon \quad (4)$$

$$m_{eff} = \rho AL\int_0^1 \Phi_i^2 d\varepsilon + \sum_j m_j\left(\Phi_i\left(\varepsilon_j\right)\right)^2 \quad (5)$$

where K_{eff} is the equivalent stiffness, m_{eff} is the equivalent mass, E is the elastic modulus of material, I is the moment of inertia of cross-section, F is the axial load on the beam, ρ is the density of material, A is the cross-sectional area, ϕ_i is i[th]-order vibration mode, L is the length of the beam, ε defined as x/L which range is 0 to 1.

Resrach supported by the National Natural Science Foundation of China (51705430 and 51775447) and the Fundamental Research Funds for the Central Universities (G2017KY0102).

All Authors are with MOE Key Laboratory of Micro and Nano Systems for Aerospace, Northwestern Polytechnical University, Xi'an, China, 710072 (*Corresponding author email*: shenq@nwpu.edu.cn).

978-1-5386-4251-1/18 $31.00 © 2018 IEEE

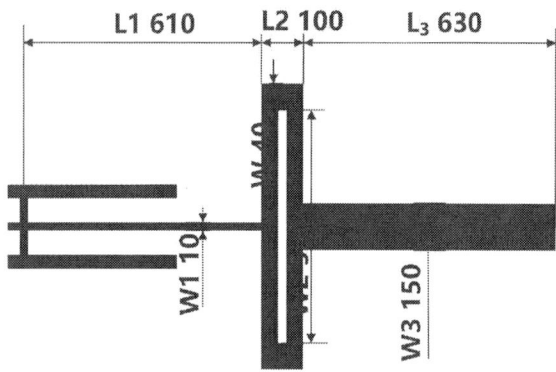

Figure.2 Simplified schematic of the resonant beam.

Theoretically, when ambient temperature is 0 degree centigrade, the calculated first frequency is 18962.7Hz. If temperature increase 10 degrees centigrade, the first frequency is 22050Hz. The theoretical sensitivity is 309Hz/K.

Fig.3 (a) implements a strain simulation based on COMSOL multiphysics: the resonant beam is loaded an axial force. Fig.2 (b) illustrate the simulation result: the frequency shift is amplified with stain amplifying effect, and the sensitivity is 263Hz/K.

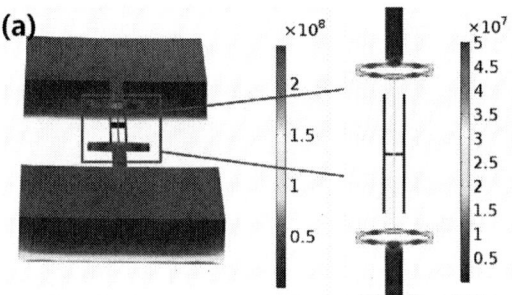

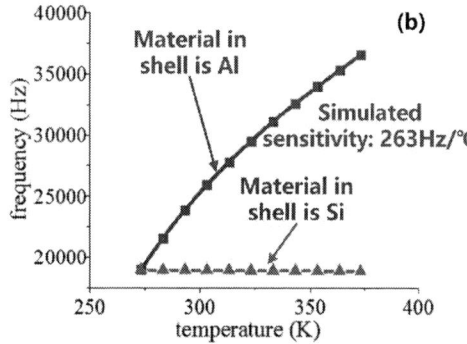

Figure.3 (a) thermal strain distribution of resonator (b) graph of frequency versus temperature with different package material.

III. FABRICATION AND EXPERIMENT

The resonator was fabricated based on SOI process, as shown in Fig.4. Fig.5 show the photograph of test circuit and the test result: the measured sensitivity is 252Hz/K.

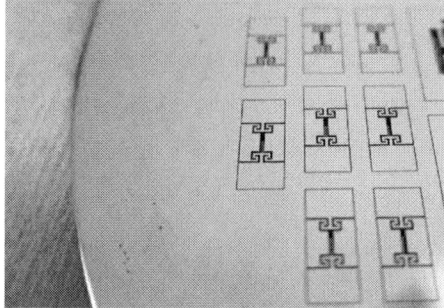

Figure.4 Picture of the fabricated resonator on wafer

(a)

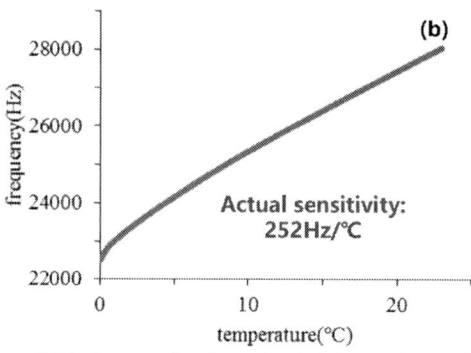

Figure.5 (a) photograph of test circuit (b) experimental test result

IV. CONCLUSION

This study presents a high-sensitivity micro resonant temperature sensor by using materials with different thermal expansion coefficients in package and resonator. A theoretical sensitivity up to 309Hz/K is calculated. The simulated sensitivity is 263Hz/K, and actual sensitivity is 252Hz/K, which closely matches theory and simulation result. The middle glass layer is voided, and a higher-sensitivity temperature sensor is achieved by the proposed strain-amplification method in this work.

REFERENCES

[1] Melamud R, Kim B, Chandorkar S A, et al. Temperature-compensated high-stability silicon resonators[J]. Applied physics letters, 2007, 90(24): 244107.

[2] Ng E J, Lee H K, Ahn C H, et al. Stability of silicon micro electro mechanical systems resonant thermometers[J]. IEEE Sensors Journal, 2013, 13(3): 987-993.

[3] Kose T, Azgin K, Akin T. Design and fabrication of a high performance resonant MEMS temperature sensor[J]. Journal of Micromechanics and Microengineering, 2016, 26(4): 045012.

Electromechanical Piezoresistive Sensing of Graphene-based Intracranial Pressure Sensor

M. Mohamad, N. Soin, *Member, IEEE,* and F. Ibrahim, *Member, IEEE*

Abstract— Graphene shows a promising future in the application of biomedical sensors as the piezoresistive sensing elements due to its electromechanical properties. This paper presents the fundamental development stage of graphene-based piezoresistive intracranial pressure sensor, i.e. to determine its diaphragm design, which is made of polydimethylsiloxane polymer. Different thicknesses of a square diaphragm were simulated using COMSOL Multiphysics. The Parametric Sweep function was used to simultaneously simulate the changes of two parameters, namely diaphragm thickness and operating pressure. It was found that the thin diaphragm is more susceptible to deform due to the rapid geometry changes and the differences in modulus of elasticity of the materials used in the design. Meanwhile, the stress experienced by the diaphragm degraded with the increase in thickness. However, a slight modification in designing and positioning the piezoresistors would make the sensor's performance on par with those of thin diaphragm. Hence, by selecting the right thickness and shape of polydimethylsiloxane diaphragm, it will serve as a good platform in developing the graphene-based piezoresistive intracranial pressure sensor.

Keywords— pressure sensor; MEMS; diaphragm

I. INTRODUCTION

The tremendous mechanical properties of graphene make it the vogue material to be researched for its potential in the sensing applications [1-3]. Moreover, its biocompatibility makes graphene a suitable material for biomedical sensing application. Graphene has been seen in the pressure sensor and strain gauge applications due to its piezoresistance effect [1-4]. It is found that the imperfection of fabricated graphene helps in improving the electromechanical properties of graphene which consequently enhances the sensing ability of the device [3]. However, there is still ambiguity in terms of the electromechanical properties of graphene [1, 5, 6].

This project explores the potential of graphene for the intracranial pressure sensing application especially in understanding the electromechanical properties of graphene. Nonetheless, this paper presents the fundamental stage of the sensor design, i.e. the diaphragm. Polydimethylsiloxane

(PDMS) polymer was chosen for the diaphragm due to its biocompatible properties. Although PDMS diaphragm can deform dynamically during operation [7], the intracranial pressure sensor is normally applied for a short duration and operated at low pressure sensing, like the sensor developed by Lee and Choi which used PDMS as a diaphragm [8].

II. SENSOR DESIGN

Fig. 1 shows the design of graphene-based intracranial pressure sensor, while the materials for the sensor are listed in Table 1. A conventional flat square diaphragm of $200~\mu m \times 200\mu m$ was chosen. The structure in Fig. 1 without the circuitry elements was simulated using the COMSOL Multiphysics software to determine the suitable thickness of PDMS diaphragm and also the high stress regions for the placement of graphene piezoresistors. The pressure was applied uniformly to the top surface while a fixed constraint was defined at the bottom structure. The simulation was run for the diaphragm thickness, h ranging from 10 μm to 20 μm and the applied pressure changes from 0 mmHg to 100 mmHg using the Parametric Sweep function. The SiO₂ layer was kept at 5 μm during the simulation.

III. RESULTS AND DISCUSSION

Fig. 2(a)-(c) shows the surface von Mises stress of the PDMS diaphragm when the applied pressure is 100 mmHg. For the top surface, the results indicate that the high stress regions are located at the center of the diaphragm and also at the middle points along the diaphragm's edges. As the thickness increases, the high stress regions along the edges have moved further into the bulk region. Meanwhile, for the bottom surface, the results show that the maximum stress for 10 μm and 15 μm thicknesses exceed the yield strength of

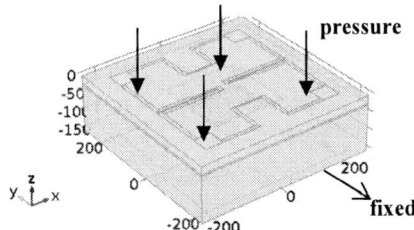

Figure 1. Structure of the graphene-based intracranial pressure sensor.

TABLE I. MATERIALS FOR THE SENSOR.

Part	Material
Substrate	n-type Si (100)
Insulator layer	SiO₂
Diaphragm	PDMS
Resistor	graphene
Electrode	Al

M. Mohamad is with the Electrical Engineering Department, Faculty of Engineering, University of Malaya, 50603 Kuala Lumpur, Malaysia and Department of Electronic Engineering, Faculty of Electrical and Electronic Engineering, Universiti Tun Hussein Onn Malaysia, 86400 Batu Pahat, Johor, Malaysia (e-mail: mazita@siswa.um.edu.my).

N. Soin is with the Electrical Engineering Department and Centre for Innovation in Medical Engineering, Faculty of Engineering, University of Malaya, 50603 Kuala Lumpur, Malaysia (corresponding author phone: +603-7967-5205; fax: +603-7967-5316; e-mail:norhayatisoin@um.edu.my).

F. Ibrahim is with the Biomedical Engineering Department and Centre for Innovation in Medical Engineering, Faculty of Engineering, University of Malaya, 50603 Kuala Lumpur, Malaysia (e-mail: fatimah@um.edu.my).

978-1-5386-4251-1/18 $31.00 © 2018 IEEE

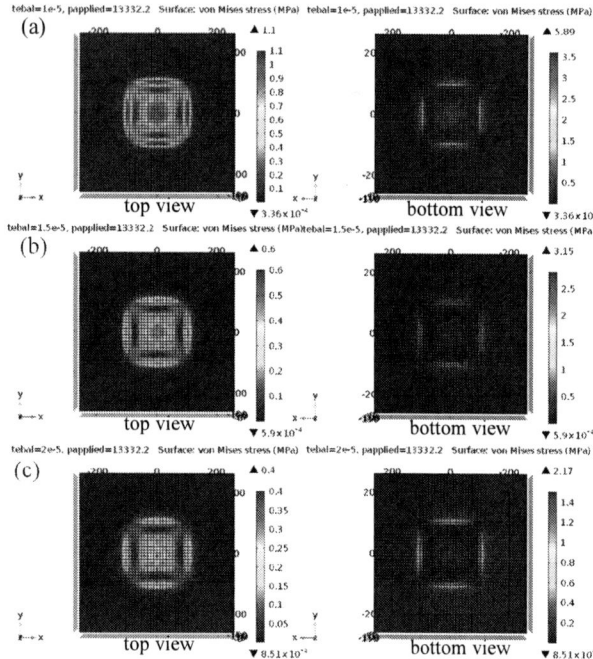

Figure 2. Surface von Mises stress of PDMS diaphragm at different thickness, h when the applied pressure is 100 mmHg (a) $h = 10$ μm; (b) $h = 15$ μm; (c) $h = 20$ μm.

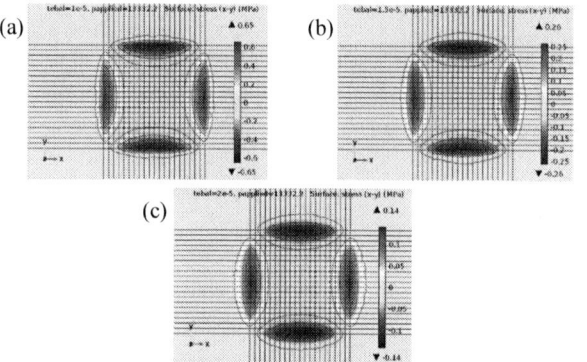

Figure 3. Distribution of stress (σ_x-σ_y) on the upper surface of PDMS diaphragm at different thickness, h when the applied pressure of 100 mmHg (a) $h = 10$ μm; (b) $h = 15$ μm; (c) $h = 20$ μm.

PDMS, i.e. about 2.2 MPa. The high stress regions are located on the diaphragm and/or near to it. This is due to the rapid geometry changes and the difference in modulus of elasticity, i.e. from PDMS layer to SiO$_2$ layer which would disrupt the flow of stresses [9]. Therefore, the diaphragm thickness of 20 μm was chosen for this sensor.

Fig. 3 shows the differential stress (σ_x-σ_y) distribution on the upper surface of the PDMS diaphragm at 100 mmHg. As the thickness increases, the high stress regions gradually shift from within the diaphragm's confinement (for $h = 10$ μm) to the bulk region (for $h = 20$ μm). Additionally, the contours of differential stress (σ_x-σ_y) distribution also become more symmetrical. Fig. 4 presents the normalized differential stress (σ_x-σ_y) on the diaphragm's top surface along the x-coordinate at 100 mmHg. It shows similar trend as [10] in which the values within the diaphragm regions are identical regardless the thickness. The changes become distinct just outside the diaphragm.

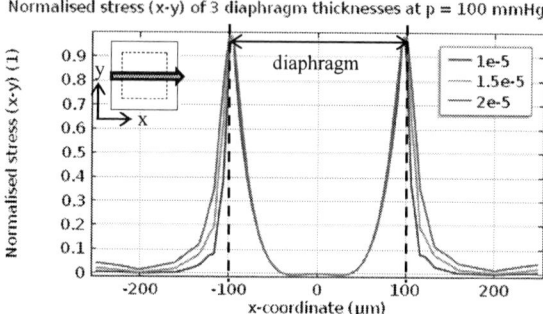

Figure 4. Normalized stress (σ_x-σ_y) distribution along the x-coordinate (y=0) on the top surface of PDMS diaphragm of different thickness, h when the applied pressure is 100 mmHg.

IV. CONCLUSION

Overall, the fundamental design stage of graphene-based piezoresistive intracranial pressure sensor, namely the diaphragm part has been accomplished. A 20 μm was found to be the right thickness for PDMS diaphragm as it would sustain during the operation. A slightly thicker PDMS diaphragm with some adjustment on the location of piezoresistors would also produce a high sensitivity sensor. By selecting the suitable thickness of PDMS diaphragm, it will provide a good platform in developing the graphene-based piezoresistive intracranial pressure sensor.

ACKNOWLEDGMENT

The authors express their deepest gratitude to the MEMS Laboratory, Faculty of Engineering, UM for facilitating the research work.

REFERENCES

[1] M. A. S. M. Haniff *et al.*, "Piezoresistive Effect in Plasma-Doping of Graphene Sheet for High-Performance Flexible Pressure Sensing Application," *ACS Applied Materials & Interfaces*, vol. 9, no. 17, pp. 15192-15201, 2017.

[2] Y. A. Samad, Y. Li, A. Schiffer, S. M. Alhassan, and K. Liao, "Graphene Foam Developed with a Novel Two-Step Technique for Low and High Strains and Pressure-Sensing Applications," *Small*, vol. 11, no. 20, pp. 2380-2385, 2015.

[3] H. Tian *et al.*, "A graphene-based resistive pressure sensor with record-high sensitivity in a wide pressure range," *Scientific reports*, vol. 5, p. 8603, 2015.

[4] S.-E. Zhu, M. Krishna Ghatkesar, C. Zhang, and G. Janssen, "Graphene based piezoresistive pressure sensor," *Applied Physics Letters*, vol. 102, no. 16, p. 161904, 2013.

[5] A. Smith *et al.*, "Electromechanical piezoresistive sensing in suspended graphene membranes," *Nano letters*, vol. 13, no. 7, pp. 3237-3242, 2013.

[6] K. Xu *et al.*, "The positive piezoconductive effect in graphene," *Nature Communications* Article vol. 6, p. 8119, 09/11/online 2015.

[7] E. Berthier, E. W. Young, and D. Beebe, "Engineers are from PDMS-land, Biologists are from Polystyrenia," *Lab on a Chip*, vol. 12, no. 7, pp. 1224-1237, 2012.

[8] D.-W. Lee and Y.-S. Choi, "A novel pressure sensor with a PDMS diaphragm," *Microelectronic Engineering*, vol. 85, no. 5, pp. 1054-1058, 2008.

[9] W. C. Young and R. G. Budynas, *Roark's formulas for stress and strain*. McGraw-Hill New York, 2002.

[10] Y. H. Zhang, C. Yang, Z. H. Zhang, H. W. Lin, L. T. Liu, and T. L. Ren, "A novel pressure microsensor with 30-μm-thick diaphragm and meander-shaped piezoresistors partially distributed on high-stress bulk silicon region," *Sensors Journal, IEEE*, vol. 7, no. 12, pp. 1742-1748, 2007.

Formatting Issue - Best Available Paper/Graphic

2018 IEEE 8th International Nanoelectronics Conference (INEC)

Piezoresistive Effect of Interdigitated Electrode Spacing Graphene-based MEMS Intracranial Pressure Sensor

Rahman S.H.A, Soin N., and Ibrahim F., *Member, IEEE*

Abstract— Two-dimensional (2D) materials have recently drawn great attention among researchers for emerging electronics. Among these materials, graphene has shown great potential in various types of sensor applications due to its superior electronic and mechanical properties. Its two-dimensionality as well as its high flexibility, conductivity, and transparency make graphene a promising candidate for flexible electronics. This paper reports the development of resistive graphene-based MEMS pressure sensor integrated with interdigitated electrode. These interdigitated electrode structure act as pressure magnifying structure as well as reducing the output non-linearity. A COMSOL simulation was carried out for design optimization of the resistive pressure sensor. In this study, the effect of optimization of the spacing between the Al electrodes is presented to improve the performance of graphene-based pressure sensors at room temperature. Three different spacing distances of 10, 20 and 40 μm were used as the experimental parameters. The increased spacing could affect in increasing tensile strain on graphene and increased defect generation at the grain boundaries. Therefore, the pressure sensor response could also be improved by increasing the spacing of the interdigitated electrode.

I. INTRODUCTION

Piezoresistive Micro electro mechanical system (MEMS) pressure sensor is a device that measure change in pressure by means of a change in resistivity of a piezoresistor placed on the maximum strain area of a diaphragm that is proportional to applied stress and, subsequently, to applied pressure. The objective of sensor design for biomedical applications is to maximize the performance of the sensor which can be quantified by the product of signal to noise (S/N) ratio and sensitivity to the temperature coefficient of piezoresistance in order to meet the desired design specification.

Researches has shown that the most influential parameter for the interdigitated piezoresistive pressure sensor design in the dimension of electrode geometry [1], [2]. Lin et al. (1999) found out, if the resistance of the sensing resistor is increased (by increasing the sheet resistance) or decreased (by increasing the width), the output voltage increases or decreases, accordingly. The same goes to Tsai et al., (1999) who has confirmed that the control factors such as the of sensor membrane thickness, number of electrode finger pairs and the electrodes are essential design parameters in that they significantly influence the precision and sensitivity of the sensor [3]. Currently, a study also proved that the optimum design parameters are contributed by the sensitive layer including thickness and coverage area [4].

As shown in Figure 1, the resistance (R0) of the sensor in the absence of pressure was defined by the natural contact between the single layer graphene. The applied force subsequently induced both a contact area increase (ΔX) and distortion of the graphene atomic structure by stretching (ΔY). This resulted in the piezoresistive effect (R = R0 + ΔR) of the sensor in two ways: positively for the contact effect, and negatively for the electromechanical effect, of graphene [5].

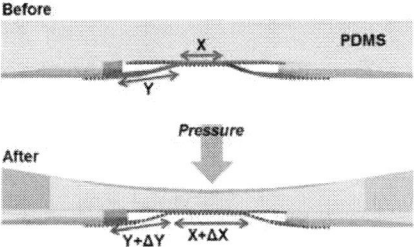

Figure 1. The sensors feature two ways of contributing sensitivity: positively for the contacting effect (X) and negatively for the electromechanical effect of graphene (Y*)*

In a typical construction of interdigitated electrode graphene-based pressure sensor, usually, flat graphene sticks well to a flexible substrate by weak van der Waals interaction. However, under a large structural deformation, sliding or even fracture is expected for graphene on its supporting substrate due to the rigid nature of graphene within its basal plane [6]. Until now, strained graphene on interdigitated electrode for pressure sensing have not been clearly investigated. Thus, the goal here is to develop an easily fabricated, high-performance graphene-based sensor capable of room-temperature operation, and, in particular, exhibiting high sensitivity and reversible fast response. For this purpose, an electrode with IDEA structure has been proposed to measure the change in resistance of strained graphene as well as to act as pressure magnifying scheme for the sensor.

II. METHODOLOGY

A. Theory

For an electrical cell comprised of two conducting terminals separated by some dielectric or conductive medium of permittivity ϵ or conductivity σ, it is possible to find the capacitance (C) or resistance (R) of the cell, respectively, via the following relations [2]:

$$= \frac{\Delta}{\sigma} \qquad (1)$$

978-1-5386-4251-1/18 $31.00 © 2018 IEEE

Formatting Issue - Best Available Paper/Graphic

$$\left| \iint \rightarrow . \rightarrow \right|$$ (2)

Here, ΔV is the voltage difference between the two terminals, $\hat{E}$ is the electric field as a function of position, and S is the surface of one of the terminals. For an ideal parallel plate capacitor (with no fringe fields) of plate area A and plate separation d, these integrals are easily evaluated when ϵ and σ are constants in space

$$= \underline{\qquad}$$ (3)

$$=$$ (4)

Therefore, if the (possibly nontrivial) cell geometry of interest can be transformed into a geometry resembling a parallel plate configuration in such a way that preserves the orthogonality of electric field lines and equipotential lines (as demanded by the solution to Laplace's equation which specifies the voltage at every point in the cell), then that transformation will easily provide the desired values of resistance and capacitance.

B. FEA analysis

The sensor consists of an interdigitated electrode transducer etched onto a flexible substrate and covered with a few layers of graphene with 0.5 μm thickness. The pressure sensor may have numbers of identical electrodes, and each electrode can be a few times longer than it is wide. Thus, the edge effects has been neglected in the simulation and the model geometry has been reduce to the periodic unit cell shown in Figure 2. This interdigitated electrode is also act as pressure amplifying design that can manipulate the electrical properties of the graphene structural properties when pressure applied in Z-direction. The simulations of the stress distribution and deflection on the graphene layer due to pressure being applied on its surface were both accomplished with the finite element analysis software, COMSOL Multiphysics. The main steps of the simulation include modelling (building geometry), material selection, setting the boundary condition (fix support and load), initial condition, meshing, and finally, running the model and viewing the results.

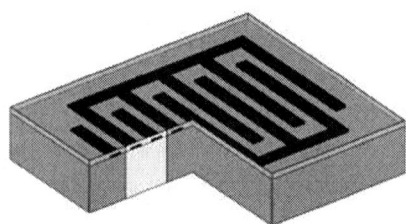

Figure 2 Geometry of the interdigitated graphene-based pressure sensor unit cell used in COMSOL model

III. RESULT & DISCUSSION

Figure 3 below shows the colour contour of a graphene IDEA deflection and strain with the applied stress simulated in COMSOL multyphysics software. Periodic boundary conditions was used to dictate that the electric potential and displacements and was found to be the same along both vertical boundaries of the geometry.

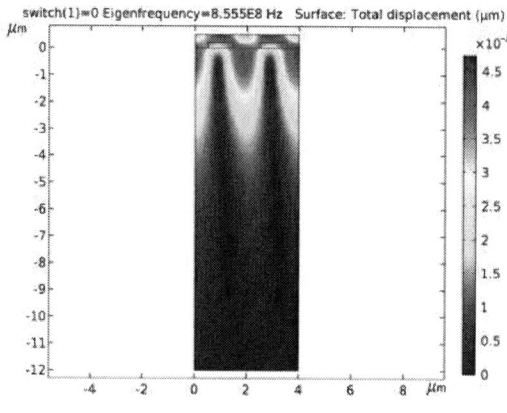

IV. CONCLUSION

A better solution in designing graphene-based pressure sensor has been suggested in this study with the application of interdigitated electrode. The integrated electrode structure act as sensing electrode as well as pressure amplifying scheme. The sensor response was significantly improved with the integration of IDEA structure.

ACKNOWLEDGMENT

This research is funded by the University Malaya research grant UMRG-AET (Grant Number RP009B-13AET), high impact research (HIR/eng/19) and flagship (FL001a-14AET). The authors would also like to express their gratitude to the microelectronic laboratory, faculty of engineering for facilitating the research

REFERENCES

[1] C. M. Yang, T. C. Chen, Y. C. Yang, M. Meyyappan, and C. S. Lai, "Enhanced acetone sensing properties of monolayer graphene at room temperature by electrode spacing effect and UV illumination," *Sensors Actuators, B Chem.*, vol. 253, pp. 77–84, 2017.

[2] K. J. Latimer, J. W. Evans, M. A. Cowell, and P. K. Wright, "Modeling of Interdigitated Electrodes and Supercapacitors with Porous Interdigitated Electrodes," *J. Electrochem. Soc.*, vol. 164, no. 4, pp. A930–A936, 2017.

[3] H. H. Tsai, D. H. Wu, T. L. Chiang, and H. H. Chen, "Robust design of SAW gas sensors by Taguchi dynamic method," *Sensors*, vol. 9, no. 3, pp. 1394–1408, 2009.

[4] H. Guo, X. Chen, and Z. Wu, "A feasible simulation method for vapor sensor based on polymer-coated NEMS diaphragm," *Measurement*, vol. 68, pp. 219–224, 2015.

[5] S. Chun, Y. Kim, H.-S. Oh, G. Bae, and W. Park, "A highly sensitive pressure sensor using a double-layered graphene structure for tactile sensing.," *Nanoscale*, vol. 7, no. 27, pp. 11652–9, 2015.

[6] Y. Wang, R. Yang, Z. Shi, L. Zhang, D. Shi, E. Wang, and G. Zhang, "Super-elastic graphene ripples for flexible strain sensors," *ACS Nano*, vol. 5, no. 5, pp. 3645–3650, 2011.

Multi-layer Noncontact Disk-shaped Electrostatic Microgenerator

B.Q. Wang, Y.X. Chen, L.H. Tang and K. Tao*, *Member, IEEE*

Abstract— In this work, we present the fabrication and characterization of a novel multi-layer noncontact disk-shaped electrostatic microgenerator. The microgenerator aims to harvest kinetic energy from rotary motion in our daily life. The multi-layer structure is composed of pairs of energy harvesting units. Each unit consists of two parts: the rotational blades with copper electrodes that are attached to the center shaft and the stationary disk that is fixed on the external barrier structure. Compared to the previous two-plate structure, the current device has two unique merits: First, both sides of the stationary disk are coated with electret material and are corona charged. It is beneficial to maximize the output power density of the whole device. Second, with the help of the micro rotary bearing, multi-layer rotary structure has been successfully implemented for the first time. Therefore, the overall performance has been multiplied by several folds and high output power can be readily achieved.

I. INTRODUCTION

Energy harvesting systems refer to the devices that are capable of scavenging and converting used environmental energy to electrical energy. Unlike the wind turbine farms and large-scale solar panels, energy harvesting systems mainly focus on low-level ambient energy and powering for low-power-consumption electronics, such as wearable electronics or wireless sensor networks. Kinetic energy is ubiquitously existed and readily available in our daily life. Kinetic energy can be transformed to electrical energy by piezoelectric, electrostatic, electromagnetic, magnetostrictive and triboelectric mechanisms [1-2]. Among these, electret-based electrostatic energy harvester is more likely to be miniaturized and used in low-frequency applications [3-6]. Electrets are dielectrics with quasi-permanent electric charge or dipole polarization, which can maintain an electric field around the structure for tens of years [3].

In the past years, several types of micro rotational power generator based on electret/electrostatic or triboelectric mechanisms have been developed [7]. However, the overall output power is limited by its two-plate structure, where only adjacent two faces of the two-plate electrodes are used for electrostatic induction or triboelectrification. In the current work, we propose a multilayer noncontact disk-shaped electrostatic microgenerator that has two merits: (a) the electret materials are coated on the both side of the stationary

This research is supported by National Natural Science Foundation of China (Grant No. 51705429) and the Fundamental Research Funds for the Central Universities. (*Corresponding author: K. Tao*)

B.Q. Wang and K. Tao are with the Ministry of Education Key Laboratory of Micro and Nano Systems for Aerospace, Northwestern Polytechnical University, Xi'an, 710072, China. (e-mail: taokai@nwpu.edu.cn;)

L.H. Tang is with Department of Mechanical Engineering, University of Auckland, 20 Symonds Street, Auckland 1010, New Zealand. (e-mail: l.tang@auckland.ac.nz)

disk and then double-sided are corona charged, which is beneficial to maximize the overall performance; (b) propose an easy-to-dissemble mechanism with barrier layer structure and micro bearings. Multilayer structure can be conveniently constructed for high output purposes.

II. DEVICE CONCEPT

The schematic of the proposed electrostatic generator is shown in Fig. 1. It consists mainly of copper blades, multi-layer disks, a barrier layer structure, a bearing and a motor. Among them, each side of every disk contains three layers: a resin supporting template, an induction copper foil and an FEP electret thin film. The barrier layer structure not only facilitates the air gap control between disks but also benefits the install and dissemble process of disks. The FEP films serve both as electret membranes and as insulating layers to prevent short circuit problem between the blade and the copper layer. The steel shaft is connected to the motor by joint coupling. During the operation, only the blades are rotated with the shaft while other multi-layer disks are kept stationary. The pattern of fan-shaped copper foils and FEP films are identical to each other to maximize the capacitance variation. When an external voltage is applied to the motor, a relative rotation is generated between blades and multi-layer disks, giving rise to induced charge circulation between two parts of the copper foils on the disks.

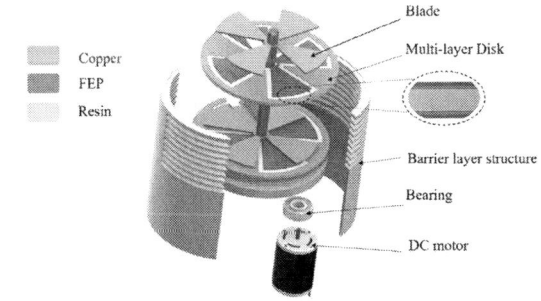

Fig.1 Schematic structure of multi-layer noncontact disk electrostatic Microgenerator

The operation principle is illustrated in Fig. 2. The generator is composed of two parts: the rotational blades and the stationary disks. The stationary part consists of two separated copper foils functioned as two electrodes but only one of them is covered by an FEP film. When a blade moves above two copper electrodes, induced charges of those electrodes will change. If we connect the two electrodes using resistance as a load, we can detect voltage between them. During a full cyclic motion of the rotational disk, the direction of current between two electrodes changes twice which generates altering signal.

978-1-5386-4251-1/18 $31.00 © 2018 IEEE

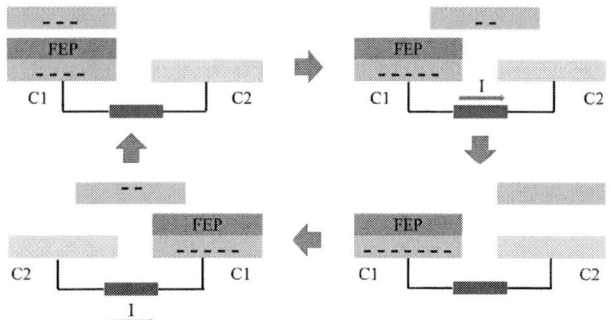

Fig.2 Schematic illustrations showing the working principle of electrostatic microgenerator

III. DEVICE FABRICATION

The whole structure can be divided into several parts. Among them, the blades and the steel shaft are fabricated by conventional mechanical machining. The barrier layer structure is made from 3D printing. Each disk is slotted for leaving the space to pass the shaft when inserted into barrier layer structure, as shown in Fig. 3. In order to decrease the thickness of the multi-layer disks, very thin tailored copper foils (10 μm) and FEP film (50 μm) were attached on the basin templates. In addition, the barrier layer structure not only supports each disk and make their surfaces parallel to each other but also make construction and disassembly process more convenient.

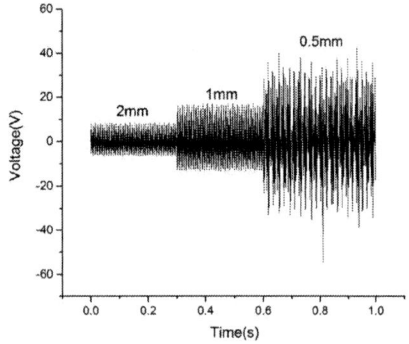

Fig.3 Photograph of the prototype: (a) fabricated multi-layer disk microgenerator; (b) the stationary disk; (c) the rotary blade

IV. TESTING RESULTS

A data acquisition system (DAQ NI USB-6289 M series) is utilized to monitor the electrical response of the microgenerator. In the first stage, a single unit with two plate rotational structure with different gap distances was investigated, as depicted in Fig. 4. It is found the output voltage drops dramatically with the gap distance increases.

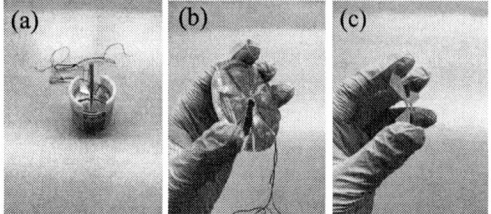

Fig.4 Electrical output voltages of the Microgenerator with different air gaps from 2mm to 0.5mm

To optimize the overall performance, the output power and voltage with different load resistances were studied. The diameter of the blade and the air distance are controlled at 6 cm and 2mm, respectively. In the experiment, we measured the output voltage by which we can also calculated the power with varied load resistances, as shown in Fig.5. In general, the voltage increases with load resistances but the current is in the opposite trend. It is seen that the maximum output power is 5.39 μW at a load resistance of 5.75 MΩ.

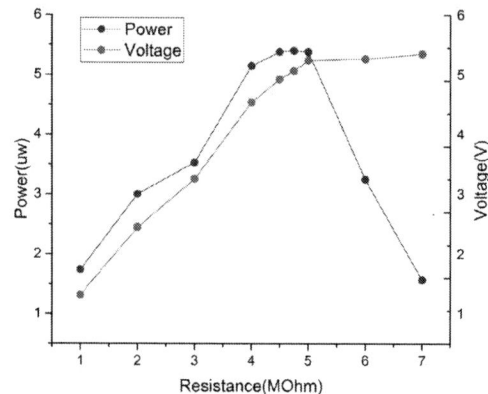

Fig.5 The measured output voltage and calculated power output across an external load with varible resistances.

V. CONCLUSIONS

In this paper, we have designed and fabricated a multi-layer noncontact disk-shaped electrostatic microgenerator for enhanced output performance. For single two-layer unit, an output power is 5.39 μW at a load resistance of 5.75 MΩ was obtained. The output performance can be further enhanced by decreasing the air gap and increasing the layers of the energy harvesting units. It is believed that the output power can achieve to mW or W level with further optimization process. The design methodology of current device could provide a convenient and effective way to achieve high performance energy harvesting devices.

REFERENCES

[1] L.H. Tang, Y. Yang, and C.K. Soh, Broadband Vibration Energy Harvesting Techniques, *Advances in Energy Harvesting Methods*, Springer, pp. 17-61, 2013.

[2] K. Tao, J. Wu, L. Tang, X. Xia, S.W. Lye, J.M. Miao and X. Hu, "A novel two degree of freedom MEMS electromagnetic vibration energy harvester", *J. Micromech. Microeng.* Vol. 26, p. 035020, 2016.

[3] Y. Suzuki, "Recent Progress in MEMS Electret Generator for Energy Harvesting", *Trans. Electr. Electron. Eng.* vol. 6, pp. 101-111, 2011.

[4] K. Tao, J. Wu, L.H. Tang, L. Hu, S.W Lye and J.M. Miao, "Enhanced electrostatic vibrational energy harvesting using integrated opposite-charged electrets", *J. Micromech. Microeng.* vol. 27, p. 044002, 2017.

[5] K. Tao, L.H. Tang, J. Wu, S.W. Lye, H.L. Chang and J.M. Miao, "Investigation of Multimodal Electret-based MEMS Energy Harvester with Impact-induced Nonlinearity", *J. Microelectromech. Syst.* pp. 1-13, 2018, DOI: 10.1109/JMEMS.2018.2792686.

[6] K. Tao, S.W. Lye, L.H. Tang, X. Xia, J.M. Miao, and X. Hu, "Out-of-plane electret-based MEMS energy harvester with the combined nonlinear effect from electrostatic force and a mechanical elastic stopper", *J. Micromech. Microeng.* Vol.10, p. 104014, 2015.

[7] J. Boland, C. Yuan-Heng, Y. Suzuki, and Y. C. Tai, "Micro electret power generator," *IEEE MEMS 2003*, pp. 538-541, 2013.

Distribution of Sn in Strained Ge$_{1-x}$Sn$_x$ (001): The Effect of Surface Passivation

Sheau Wei Ong and Eng Soon Tok [*]

Electronic Materials Growth and Interface Characterisation (ƐMaGIC) Lab,
Department of Physics, National University of Singapore, 2 Science Drive 3, Singapore 117551, Singapore.

[*]Phone: +65-6516-1192 Email: tokes@nus.edu.sg

Abstract

The effect of surface passivation on tin distribution in Ge$_{1-x}$Sn$_x$(001)/Ge(001) are studied using first principles calculations. The segregation of Sn atoms towards the surface were suppressed when the clean surface is fully passivated with hydrogen adatoms while changing the passivating species to halogen adatoms resulted in enhancing Sn segregation towards the surface. This effect strengthens when moving down the Group-VII elements from fluorine to iodine. For both hydrogenated and halogenated surfaces, aggregation of sn atoms is not favored.

1. Introduction

Incorporation of Tin in Germanium system forming Ge$_{1-x}$Sn$_x$ improves not only the carrier mobility [1-4] but also changes its band structure from indirect to direct bandgap thus attracting attention to its potential use in applications in the electronic and optoelectronic industries [5-8]. However, during growth processes such as CVD, the low miscibility of Sn in Ge leads to phase separation and formation of islands [9].

The stability of tbe Ge$_{1-x}$Sn$_x$ film can be improved by applying a passivation layer using adatoms such as H and S [10]. H adatoms added in CVD processes is known to help tin distribution into the bulk giving rise to better crystallinity of the film [11]. Both MBE growth and DFT calculations also confirms this Sn distribution.[12,13] However, for halogen adatoms such as Cl which is produced as a side product in CVD processes, the effect is not as well studied. The study of the effects of these passivation species on Sn distribution and agglomeration will contribute to its applications for electronic and optoelectronic devices.

In this work, we have therefore performed DFT calculations to examine how hydrogenation(H) and halogenation (F and I) on the clean surface will affect the Sn segregation, agglomeration and interlayer distribution in a Ge$_{1-x}$Sn$_x$ system containing 1.25% and 2.5% of Sn.

2. Methodology

Spin-polarized calculations were carried out using the plane-wave density functional theory problem PWSCF (Quantum Espresso Version 5.4.0) [14]. For Ge, Sn, H,F and I, thc Projector Augmented Wave [15] pseudopotential was used with Perdew-Burke-Ernzerhof generalized gradient approximation (PBE-GGA) for the exchange-correlation functional [16]. A supercell with dimension 8.0A° x 16.0A° x 28.0A° which consisted of 10 atomic layers of Ge or Sn was used in all calculations. The bottom layer of Ge was saturated with hydrogen atoms and this layer of hydrogen together with the bottom two layers of Ge were fixed in position to mimic a bulk-like structure. 40Ry and 480Ry were used as the energy and density cutoff respectively and the Monkhorst-Pack grid is 8x4x1. Two different Sn compositions of 1.25% and 2.5% corresponding to respectively 1 and 2 Sn atoms in the Ge-Sn system were studied.

3. Results and Discussion

Figure 1 shows the most stable configurations for 1Sn and 2Sn system. For unpassivated 1.25% Sn system, the single Sn atom is most stable when it is located in the top layer occupying the buckled up position as shown in Fig. 1a1. When there are two Sn atoms in the system, the Sn atoms prefer to be in the top layer of the system, on separate dimers and both in the buckled up position (Fig. 1b1) With H passivation, the single Sn atom prefers to be in the 3rd Ge layer (3La) of the system (Fig. 1a2). For 2.5% Sn, 1 Sn atom to be at the top layer while a second Sn occupies the 3rd Ge layer (1Sn1L2Sn3Lc) as shown in Fig. 1b2. In all cases of halogenation, the single Sn atom prefers to be at the top layer (1L) with Sn being buckled up (Fig 1a3). Halogen passivation also prefers the Sn atoms to occupy the top layer (1L). When both Sn atoms are on the surface, the Sn atoms prefer to form dimers (1Sn2Sn) when the surface is fully fluorinated (Fig 1b3) while forming mixed dimers (1Sn3Sn) when the surface passivation is by I (Fig 1b4). The relative energies for the configurations of Sn in the first four layers for 1.25% Sn and selected most stable configurations for 2.5% Sn are summarized in Table 1 and 2 respectively. The passivation energy E$_{H/X}$ for the most stable passivated configuration are calculated and shown in Figure 1. It is found that using diatomic molecules, F$_2$ is the most reactive and reactivity decreases down the group and all halogens are more reactive than molecular H$_2$. It should be noted that if radical atoms are used, F is still the most reactive while H is more reactive than I.

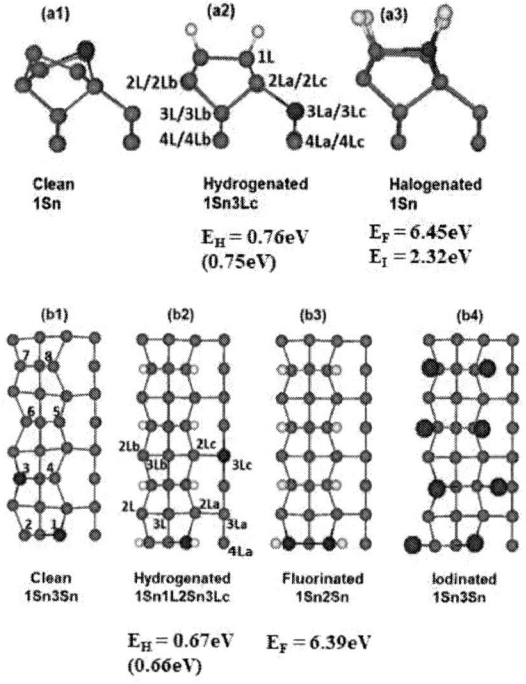

$$E_{H/X} = E_{passivated} - (E_{clean} + 4*molecule\ H_2/X_2)$$

Fig. 1a1-a3 Side view of the most stable configurations for the clean, hydrogenated and halogenated (F and I) 1.25% Sn configurations. Fig. 1b1-b4 Side view and top view for the most stable clean, hydrogenated, fluorinated and iodinated 2.5%Sn composition. Blue atoms represent Sn atoms, grey atoms represent Ge atoms, white atoms represent H, light blue atoms represent F and purple atoms represent I. Values in bracket are for most stable Sn structure in 1L.

Configuration	Passivation			
	clean	H	F	I
1Sn	0.00	0.06	0.00	0.00
1Sn2L	0.28	0.09	0.21	0.36
1Sn3L	0.37	0.27	0.25	0.55
1Sn3Lc	0.28	0.00	0.11	0.35

Table 1: Relative energies of different sublayer configurations for clean and fully passivated Sn (1.25%)/Ge(001) system.

Configuration	Passivation			
	clean	H	F	I
1Sn2Sn	0.04	0.04	0.00	0.30
1Sn3Sn	0.00	0.06	0.02	0.00
1Sn2Sn2L	0.60	0.17	0.52	0.90
1Sn2Sn3L	0.58	0.15	0.32	0.80
1Sn1L2Sn2Lb	0.28	0.09	0.23	0.31
1Sn1L2Sn3Lc	0.28	0.00	0.13	0.27

Table 2: Relative energies of selected sublayer configurations for clean and fully passivated Sn (2.5%)/Ge(001) system.

4. Conclusions

The impact of hydrogen and halogen adsorption for Sn composition of 1.25% and 2.5% for a $Ge_{1-x}Sn_x(100)$ surface has been investigated using plane-wave density functional theory. The results show that when the surface is fully hydrogenated, Sn prefers to be distributed in the sub-surface layers. For full fluorination, Sn atoms prefer to form pure dimers on the surface layer while I prefer mixed dimers. While fluorination reduces the energy for interlayer mixing compared to clean surface, iodination increases the energy. Thus during CVD growth using hydride and fluoride precursors, Sn segregation to the surface will be suppressed in a $Ge_{1-x}Sn_x$ epitaxial thin film in the hydrogen/halogen limited growth regime. On the other hand, I precursors will enhance surface segregation. Therefore the careful usage of hydrogenation, halogenation or mixture of both offers possibility to manipulate the position of Sn atoms in $Ge_{1-x}Sn_x$ system.

Acknowledgements

The authors acknowledge Grant Support from MOE AcRF : R-144-000-388-114. The computational work for this article was performed on resources of the National Supercomputing Centre, Singapore (https://www.nscc.sg) and of HPC NUS.

References

[1] X. Gong et al, IEEE Electron Device Lett. **34** (2013) 339
[2] S. Gupta et al., MRS Bull. **39** (2014) 678
[3] X. Gong et al., ECS Trans. **64** (2014) 851
[4] P.F. Guo et al., ECS J. Solid State Sci. Technol. **3** (2014)
[5] J. Mathews et al., Appl. Phys. Lett. **97** (2010) 221912
[6] H. Lin et al., Appl. Phys. Lett. **100** (2012) 102109
[7] W.J. Yin et al., Phys. Rev. B **78** (2008) 161203
[8] V.R. D'Costa et al., Phys. Rev. B **73** (2006) 125207
[9] J.D. Gallagher et al., Appl. Phys. Lett. **105** (2014) 142102
[10] A Dimoulas et al., Advanced Gate Stacks for High-Mobility Semiconductor, Springer Berlin Heidelberg (2007) p73-111
[11] K Suda et al., J. Crys. Growth **468** (2017) 605
[12] T Asano et al., Jap. J Appl. Phys. **54** (2015) 059202 MBE
[13] H Johll et al., J. Appl. Phys **117** (2015) 205302
[14] P. Giannozzi et al. J. Phys. Condens Matter **21** (2009) 395502
[15] G. Kresse et al., Phys. Rev. B **59** (1999) 1758
[16] J. P. Perdew, et al., Phys. Rev. Lett **77** (1996) 3865

Concept, methodologies and tools for carbon for sensing devices

M. Giorcelli[1], P.Savi[2], A.Tagliaferro[1]

Abstract — **Carbon is playing an increasingly prominent role as a sensing material. The various steps that transform a raw material in a sensing device are briefly presented and discussed. The discussion then deals with the role of functionalization and the different routes to achieve it. Finally, some example of sensing applications in various fields are presented.**

I. INTRODUCTION

The development of methods and devices for identification of specific markers is one of the major challenges in the sensor field. As well-know, carbon is playing an increasingly prominent role as a sensing material [1-3]. Its high charge transfer capability and electron transport properties are some of the aspects for its outstanding transduction characteristics. Carbon materials are commonly used for both bio [4] and non-bio sensors, like for example gas sensors for electronic nose [5]. This because carbon is a rather versatile element that can be functionalized in order to be selective to a specific target.

II. CARBON MATERIALS

The various kinds of available carbon materials can be attributed to three categories: (i) conventional carbon materials (including graphite blocks, the family of carbon blacks, activated carbons and diamond), (ii) nanotextured and (iii) nanosized carbons. Nanotextured carbons comprise a wide range of carbon structures including carbon fibres, glass-like carbons, pyrolytic carbons and diamond-like carbons. Fullerenes, carbon nanotubes (CNT) and graphene are examples of nanosized carbons (or nanocarbons). Different properties of carbon material could be used to create a sensor. Specific surface is one of the properties to take into account. Greater surface means higher probability to detect targeted particles. Having a large specific surface nanocarbons are of great interest for sensor applications.

A. Synthesis and preparation

Carbon materials can be syntethised in different ways, including high temperature high-pressure synthesis [6]. For nanostructured carbon materials production, Chemical Vapor deposition (CVD) is one of the most used procedure [7]. This temperature treatment in usually inert atmosphere (N_2, Ar) is able to product different types of carbon with the presence of specific catalyst and carbon source. Another example is activated carbon that is obtained from carbonaceous precursors, such as coal, wood, nutshells, etc. using for instance thermal pyrolysis [8]. Nanotextured carbons, like carbon fibers, are produced from polymer precursors such as polyacrylonitrile (PAN) by two steps carbonization to drive off non-carbon contents. Carbon sensing devices are more oriented to use nano and micro size carbon particles conveniently functionalized.

B. Dispersion of carbon materials in polymer matrices

The dispersion of carbon fillers inside a matrix is strongly depending on the chemical "compatibility" at the interface between the carbon nanophase and polymers. Dispersion is a crucial point in order to give uniform sensing properties to the device. In general, carbon materials display poor solubility in aqueous solvents where they are prone to aggregation due to hydrophobic interactions. This effect is particularly evident in nano carbon size dimensions. However it can be addressed through surface functionalization by covalent or noncovalent methods, as it will be briefly discussed below. Surface modification is also used to introduce receptor sites for sensing. See for example ref.6, focusing on the CNT case.

III. FUNCTIONALIZATION OF CARBON SURFACES

Conventionally, a sensor is composed of a recognition element, which defines the selectivity, a transducer, which transduces the recognition event into a signal with a given sensitivity, and a detector, which allows the reading of the target detection. The ultimate goal is to develop a highly sensitive and selective sensor in a reliable manner for fast in-situ measurements of analytes. The functionalization process plays a crucial role. In this respect carbon materials have critical advantages: they can be easily functionalized, they have high charge transfer capability and good electronic transport properties.

A. Enhancing the surface sensitivity

We can distinguish two main functionalization processes: (i) the functionalization of carbon material inside a dispersion followed by deposition on a solid substrate, or (ii) the growth of carbon structure followed by in-situ functionalization. The selection of the appropriate approach depends on the recognition elements and the type of the device.

B. Covalent and non-covalent functionalization

Functionalization can be either covalent or non-covalent. The covalent approaches can be grouped into two main categories: (i) defect site chemistry and (ii) addition chemistry. The first group involves an initial oxidation process while the second one employs highly reactive species that are able to modify from sp2 to sp3 the hybridization of carbon atoms. The popular amide reaction and the diazonium coupling alternative are the main covalent reactions employed in sensing. On their side, the non-

---

M Giorcelli and A. Tagliaferro are with Department of Applied Science and Technologies, Politecnico di Torino, C.so Duca degli Abruzzi 24, 10129 Torino, Italy (email: mauro.giorcelli@polito.it; alberto.tagliaferro@polito.it)

P. Savi is with Department of Electronic and Telecommunication, Politecnico di Torino, C.so Duca degli Abruzzi 24, 10129 Torino, Italy (email: patrizia.savi@polito.it)

covalent approaches anchor receptors via supramolecular interactions. Pyrene, polymers, biomolecules and metallic nanoparticle-based are the most attractive approaches as discussed in [1].

IV. EXAMPLE OF APPLICATIONS

The final device is made by two components: sensor and transducer. The sensor is used to collect information and the transducer had to transform this information into electrical signals to be measured and recorded. Transduction mechanisms include (but are not limited to) electrical, mechanical, electromagnetic (including visible light), chemical, acoustic or thermal detection. In literature a myriad of carbon based devices are reported (see e.g. [10-15]). Some examples gathered by application area are reported below.

A. Gas sensor sensors

Gas sensor devices are present in many research fields, including environmental, biomedical and industrial. They must be easily configurable, fast responding, and with good reproducibility and sensitivity. The state of art reports different sensing and transducing strategies: electrochemical, optical, conductometric etc., based on specific chemically interactive materials. Gas sensors based on CNT are of particular interest because of their large surface to volume ratio [2]. Morover the presence of defects and the porous structure of CNTs provide effective adsorption sites for gas molecules. The gas molecules adsorbed on CNT surface modify their properties such as resistance and capacitance leading to devices very sensitive to the environment.

B. Optical based sensors

In optical based sensors, optical mechanisms such as fluorescence quenching due to the presence of target molecules, lead to a loss of emission due to the interaction between the carbon material and the analyte. This principle is at the base of the sensor mechanism. One example of a fullerene-based optical sensor for the selective detection of F−ions was recently reported in [10]. Briefly, the presence of F−ion results in a redshift of the fullerene absorption peak from 717 nm to 823 nm.

C. Electrochemical biosensors

Recent advances in electrochemical biosensor applications are made using CNTs and graphene. The key advantage of these carbon nanomaterials are their prominent electro catalytic activity towards small as well as big molecules, the lowering of redox potentials, the resistance to surface fouling and the high electroactive area. Their nano-size structure can promote the direct electron transfer of various enzymes without affecting the macro biomolecule activity over time [1].

D. Piezo resistivity sensors

Piezoresistivity occurs when an electrical resistor change its resistance when strain and deformation due to an external mechanical input are present. Sensing application using carbon based materials able to use this effect are for example accelerometers, pressure sensors, gyrorotation rate sensor, tactile sensors, flow sensors, sensors for monitoring structural integrity of mechanical elements and others. An example is reported in [16], where they illustrate the piezo resistive effect of pristine CNT thin films using a three-point bending test.

V. CONCLUSIONS

A brief non exhaustive overview of the use of carbon materials as sensors is addressed. Material versatility leads to the possibility to design new applications for future devices.

REFERENCES

[1] D. Demarchi, A. Tagliaferro "Carbon for sensing devices" *Springer International Publishing* , 2015.

[2] I. V. Zaporotskova, N. P. Boroznina, Yuri N. Parkhomenko, Lev V. Kozhitov "Carbon nanotubes: Sensor properties. A review", *Modern Electronic Materials* vol. 2 pp. 95–105, 2016.

[3] F. Baptista, S. A. Belhout, S. Giordani and S. J. Quinn "Recent Developments in Carbon Nanomaterial Sensors", *Chem. Soc. Rev.*, vol. 44, pp. 4433-4453, 2015

[4] Z. Zhu, L. G.-Gancedo, A. J. Flewitt, H. Xie, F. Moussy, W. I. Milne, "A Critical Review of Glucose Biosensors Based on Carbon Nanomaterials: Carbon Nanotubes and Graphene," *Sensors*, vol. 12, pp. 5996-6022, 2012.

[5] N. L. W. Septiani, B. Yuliarto "Review – the development of gas sensor based on carbon nanotubes," *Journal Electrochem. Soc.*, vol. 163, no. 3, pp. 97-106, 2016.

[6] A. Bazargan, Y. Yan, C. W. Hui, G. McKay "A review: synthesis of carbon based nano and micro materials by high temperature and high pressure," *Ind. Eng. Chem. Res.*, vol. 52, no. 36, pp 12689-12702, 2013.

[7] S. Porro, S. Musso, M. Giorcelli, A. Tagliaferro "Optimization of a thermal – CVD system for carbon nanotube growth," *Physica E Low-dimensional System and nanostructures*, vol. 37, no. 1, pp 16-20, 2007.

[8] M. Noman, A. Sanginario, P. Jagdale, A. Tagliaferro "Activated carbonized pistachio nut shells for electrochemical luminescence detection," *Journal of applied electrochemistry*, vol. 45 n. 6 pp. 585-590, 2015.

[9] V. Biju "Chemical modifications and bioconjugate reactions of nanomaterilas for sensing, imaging, drug delivery and therapy," *Chem. Soc. Rev.*, vol. 43, pp. 744-764, 2014.

[10] L. Xu, S. Liang, L. Gan "A green fullerene derivative as a fluoride ion sensor," *Org. Chem. Front.*, vol. 1, pp. 652-656, 2014.

[11] M. Giorcelli, P. Savi, A. Delogu, M. Miscuglio, M.H. Yahya, A. Tagliaferro, "Microwave absorbtion properties in epoxy resin multiwalled carbon nanotubes composite," *International Conference on Electromagnetics in Advanced Applications (ICEAA13)*, Torino, September 9-13, pp. 1139-1141, 2013.

[12] M. Miscuglio, M. Hajj Yahya, P. Savi, M. Giorcelli, A. Tagliaferro, "RF Characterization of polymer multi-walled carbon nanotube composites," *IEEE Conference on Antenna measurements & Applications (CAMA)*, Antibes Juan-les-Pins, France, November 16-19, pp. 1-4, 2014.

[13] M. Giorcelli, P. Savi, M. Miscuglio, M. Hajj Yahya, A. Tagliaferro, "Analysis of MWCNT/epoxy composites at microwave frequency: reproducibility investigation," *Nanoscale Research Letters*, vol. 9, no. 168, pp. 1-5, 2014.

[14] P. Savi, M. Miscuglio, M. Giorcelli, A. Tagliaferro, "Analysis of microwave absorbing properties of Epoxy MWCNT composites," *Progress in Electromagnetics Research Letters*, vol. 44, pp. 63-69, 2014.

[15] M. Giorcelli, P. Savi, M. Yasir, M. Hajj Yahya, A. Tagliaferro, "Investigation of Epoxy Resin/MWCNT composites behaviour at low frequency, " *Journal of Material Research*, Soft Nanomaterials Focus Issue, vol. 30, nr. 1, pp. 101-107, January, 2015.

[16] Y. Li, W. Wang, K. Liao, C. Hu, Z. Huang, Q. Feng, "Piezoresistive effect in carbon nanotubes films", Chinese Science Bulletin, vol. 48, n.2, pp. 125-127, 2003.

Tuning endotaxial growth of CoSi$_2$ nanowires and nanodots*

Bin Leong Ong and Eng Soon Tok

Abstract— The shape transition of the CoSi$_2$ islands from nanowire to nanodot and vice versa can be controlled by using different growth temperatures. High growth temperatures favor the formation of ridge nanowires and flat square-nanodots. At lower growth temperatures, the nanowires become more dot-like while flat nanodots are more wire-like. The islands' length, width and height follow the Arrhenius relation with activation energies ranging from 0.4 - 1.6 eV. The shape-transition of these nanowires and nanodots are kinetically limited by thermally-activated processes.

I. INTRODUCTION

There are recent interests in using cobalt disilicide (CoSi$_2$)/Si-substrate as a template for growth of spin-based electronics using Co-Heusler alloys [1-3]. The growth of CoSi$_2$ on Si with smooth surface morphology and abrupt interface is a pre-requisite to achieving good device performance. However, the growth process is complicated by endotaxial formation of 3-dimensional Type A flat islands (where CoSi$_2$(001)//Si(001) and CoSi$_2$[110]//Si[110]) and Type B ridge islands (where CoSi$_2$(221)//Si(001) and CoSi$_2$[$\bar{1}$10]//Si[$\bar{1}$10]) [4] that makes the surface morphology rough and the interface heterogeneous. In addition, these islands also undergo complex island-shape transition depending on the growth temperature. For instance, shape transition of CoSi$_2$ from nanodot to nanowire [5,6] was reported to follow the Tersoff's thermodynamically driven strain-relaxation growth model [7]. In contrast, works by Goldfarb *et al.*, Bennett *et al.*, Ong *et al.*, Scheuch *et al.*, and Adams *et al.* [4,8-12] suggest that the growth occurs endotaxially and are determined by kinetics considerations instead. Achieving the desired template morphology (flat, ridge, nanowires and nanodots) would require fundamental insights into the growth dynamics [13-16] i.e., the considerations of energetics and/or kinetics.

In this work, we therefore investigate the growth dynamics and kinetics of these nanostructures between 500°C and 800 °C at Co coverage up to 0.5 ML. We show how kinetic limiting processes influence the morphology of CoSi$_2$ nanowires on the Si(001) surface in terms of its geometries and structures for both types of islands.

*Grant Support from MOE AcRF: R-144-000-388-114.

B.L. Ong is with the Singapore Synchrotron Light Source, National University of Singapore, 5 Research Link, Singapore 117603, Singapore, and Electronic Materials Growth and Interface Characterisation (ƐMaGIC) Lab, Department of Physics, National University of Singapore, 2 Science Drive 3, Singapore 117551, Singapore

E. S. Tok is with the Electronic Materials Growth and Interface Characterisation (ƐMaGIC) Lab, Department of Physics, National University of Singapore, 2 Science Drive 3, Singapore 117551, Singapore (Corresponding author: phone: +65-6516-1192; e-mail: tokes@nus.edu.sg).

II. METHODOLOGY

The STM experiments were performed using the OMICRON Variable-Temperature Scanning Tunneling Microscope (VT-STM) *in-situ* in an OMICRON Ultra-High Vacuum (UHV) system of base pressure 2×10^{-10} mbar. Sample substrates were cut from Boron-doped p-type singular Si(001) wafers (Virginia Semiconductors) with resistivity less than 0.1 Ω.cm^{-1}. These samples were chemically etched *ex-situ* based on the recipe described by Ong *et al.* [17,18]. They were finally dipped in dilute aqueous hydrofluoric (HF) acid solution prior to outgassing in the UHV chamber for 8 hr at 300 °C. The samples were then gradually heated to 700 °C and flashed to 1100–1150 °C for 5 cycles with 30 s per cycle. Upon cooling to room temperature, the clean Si(001) surface morphology of the samples was verified *in-situ* using the VT-STM. The samples were then deposited at elevated temperatures between 500 °C and 800 °C with cobalt (Goodfellow, purity > 99.99%) using electron-beam evaporation with a rate of 0.1 ML/min for 1 min. Temperatures and deposition rates were determined using an infrared pyrometer and a quartz crystal monitor (QCM). After cooling to room temperature, the samples' surface morphologies were then characterized *in-situ* using the VT-STM.

The STM images were acquired using constant-current mode, with a tunneling current at -1.0 nA and sample bias of -2.0 V. The STM scan direction was kept unidirectional from bottom to top. The acquired images were background-corrected (planar) and flattened ("Flatten discarding regions") using the WSxM software (Nanotec Electronica) [19]. Analyses and measurements from these images were done using the same software.

III. RESULTS AND DISCUSSIONS

Fig. 1 shows the overall phase diagram for the evolution of CoSi$_2$ ridge and flat-type islands, grown at temperatures between 500 °C and 800 °C. The length of flat islands increased initially from 560 °C to 650 °C forming short nanowires. Above 650 °C, the island width increases more dramatically than its length resulting in the formation of larger but more dot-like islands. Ridge islands, on the other hand, formed short dot-like nanowires at 560 °C. The island's length is found to increase more significantly than its width with increasing growth temperatures. Consequently, the island becomes more wire-like above 650 °C.

By measuring the islands' length and width for each growth temperature and Co coverage, we observed that the increase in the islands' dimensions with respect to the growth temperature follows the Arrhenius-like relation, suggesting that the growth processes in the shape-transition of ridge-type

CoSi$_2$ nanowires and flat-type nanodots are thermally activated (see Figure 2). The results show that below 800 °C, the growth of ridge and flat type CoSi$_2$ islands is kinetically constrained [20].

CoSi$_2$ islands

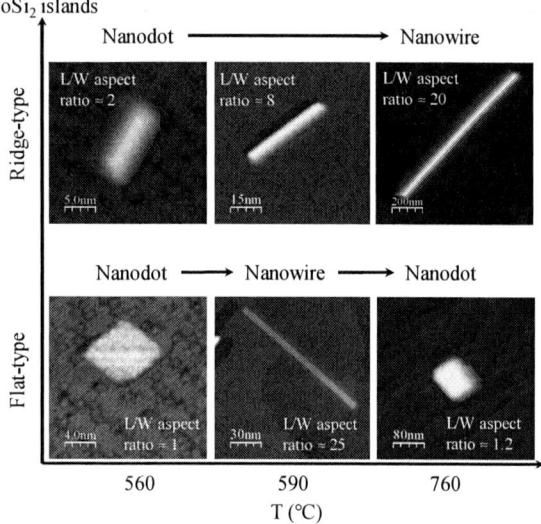

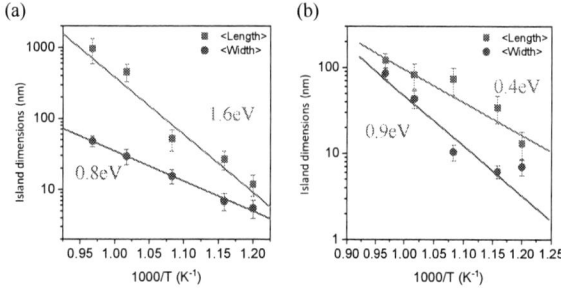

Figure 1.   Overall phase diagram of CoSi$_2$ islands as function of growth temperature

Figure 2.   Average island dimensions (length and width) of (a) ridge, and (b) flat type CoSi$_2$ islands as function of 1/T. The values indicate the activation energy barriers for each island dimension.

For ridge-type islands, the activation energy barrier is larger by 0.8 eV for the island's length compared to its width. As such, the ridge islands are shorter at low temperature. Increasing the growth temperature allows the island to become more wire-like. For flat islands, on the other hand, they show a smaller activation energy barrier compared to its width. They are therefore more compact than the ridge islands. As temperature increases to 650 °C, similarly to the ridge islands, they also evolve initially into nanowires. However, the smaller difference in the energy barriers (0.5 eV) between the flat island's length and width, compared to the ridge islands, results in the transformation of the flat nanowire into a nanodot as growth temperature increases beyond 650 °C. In contrast, the ridge islands remain wire-like. This difference in growth behavior for flat and ridge islands can be attributed to the formation of different CoSi$_2$/Si interfaces [4].

IV. CONCLUSION

We have shown in this work that the morphology of CoSi$_2$ islands (nanowires and nanodots) on Si can be tuned by changing the growth temperature. High growth temperatures favor the formation of ridge nanowires and flat square-nanodots. The shape-transition of nanodots/nanowires is thermally activated and is governed by an interplay between thermodynamics and kinetics associated with the presence of different interfacial planes, island-edge diffusion anisotropy and corner energy-barriers.

ACKNOWLEDGMENT

The authors acknowledge Grant Support from MOE AcRF: R-144-000-388-114.

REFERENCES

[1] Z. Nedelkoski *et al.*, J. Phys.: Condens. Matter **28** (2016) 395003

[2] Kuerbanjiang *et al.*, Appl. Phys. Lett. **108**, (2016) 172412

[3] A. Pokhrel *et al.*, "Growth of Metal Silicide nanowires and Their Spintronic and Renewable Energy Applications" in *Semiconductor Nanowires: From Next-Generation Electronics to Sustainable Energy*, Lu *et al.* Eds. The Royal Society of Chemistry (2015)

[4] B.L. Ong *et al.*, Surf. Sci. **606** (2012) 1649

[5] S.H. Brongersma *et al.*, Phys. Rev. Lett., **80** (1998), 3795

[6] J.C. Mahato, *et al.*, Appl. Phys. Lett., **100** (2012) 263117

[7] J. Tersoff *et al.*, Phys. Rev. Lett., **70** (1993), 2782

[8] I. Goldfarb *et al.*, Phys. Rev. B, **60** (1999), 4800

[9] P.A. Bennett *et al.*, Thin Solid Films, **519** (2011), 8434

[10] Z. He *et al.*, Phys. Rev. Lett., **93** (2004), 256102

[11] V. Scheuch *et al.*, Surf. Sci., **372** (1997) 71

[12] D.P. Adams *et al.*, J. Appl. Phys., **76** (1994), 5190

[13] R.M. Tromp and J.B. Hannon, Surf. Rev. Lett., **9** (2002) 1565

[14] J.A. Venables, Surf. Sci. **299-300** (1994) 798

[15] A. Pimpinelli *et al.*, *Physics of Crystal Growth*, Cambridge University Press (1998)

[16] Y.W. Mo *et al.*, Surf. Sci., **268** (1992), 275

[17] W.J. Ong *et al.*, Phys. Chem. Chem. Phys., **9** (2007) 991

[18] W. Ong *et al.*, Phys. Rev. B, **79** (2009) 235439

[19] I. Horcas *et al.*, Rev. Sci. Instrum., **78** (2007) 013705

[20] B.L. Ong *et al.*, Surf. Sci. **647** (2016) 84

An Improved SOI Resonant Pressure Sensor using Atmospheric Packaging*

First Sen Ren, Second Jianbing Xie, Third Qiang Shen, Fourth Fei Wang, Fifth Weizheng Yuan, and
Sixth Jinkuan Zhang

Abstract— An improved atmospheric packaged SOI resonant pressure sensor is presented. A special anchor structure using suspended connecting truss is developed to suppress the vertical position shift of the resonator when the diaphragm deflects, and a stress isolating structure is introduced to improve the performance of the resonant pressure sensor. Experimental results show that the vertical position shift of the resonator is reduced to only 7.3% compared with conventional anchor design. Over the full scale pressure range of 3.5–280 kPa, the pressure sensitivity is 10.86 Hz/kPa, with the nonlinearity is 0.0138%FS, the hysteresis error is 0.0047%FS, the repeatability error is 0.0071%FS, and the accuracy is better than 0.02%FS.

I. INTRODUCTION

Resonant pressure sensors are the optimum selection of pressure monitoring in precision measurement, for their high accuracy and extremely long-term stability [1]. They have a lot of advantages, such quasi-digital signal output and strong anti-interference. As a result, resonant pressure sensors have a very wide range of application and a huge market in military and civilian fields. In this paper, an improvement of the atmospheric packaged SOI resonant pressure sensor is proposed [2-3]. In order to suppress the vertical position shift of the resonator when the diaphragm deflects, a special anchor structure is developed. Furthermore, a stress isolating structure is introduced to improve the accuracy of the resonant pressure sensor. Experimental results demonstrate that the performance of the atmospheric packaged SOI resonant pressure sensor has been improved.

II. IMPROVEMENT OF SENSOR DESIGN

The conventional anchor structure of a resonant pressure sensor is illustrated in Fig. 1(a), in which the resonator will rise several micrometers while the diaphragm deflects under large pressure. As a result, the driving and detection capacitances will be reduced, which will increase the difficulty of closed-loop control. In order to suppress the vertical position shift of the resonator, a special anchor

structure using suspended connecting truss is developed, as shown in Fig. 1(b). When the diaphragm deflects under applied pressure, the pedestals go up, but the suspended connecting trusses go down relative to the pedestals as the rotation of the pedestals. As a result, the vertical position shift of the resonator can be inhibited.

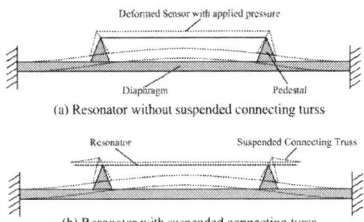

Figure 1. Comparison of two anchor structures

In order to improve the accuracy of the resonant pressure sensor, the hysteresis error, the repeatability error and the temperature coefficient must be reduced. For this reason, a stress isolating structure is introduced. Considering the thermal expansion coefficient and the mechanical strength, Pyrex 7740 glass has been chosen for stress isolation. Fig. 2 shows the axial stress of the resonator with different thickness of Pyrex 7740 glass substrates while the temperature changes from 20 °C to 60 °C. The axial stress induced from the thermal expansion coefficient mismatch decreases as the thickness of the glass substrate increases. Therefore, thicker glass substrate is fit for reducing temperature coefficient of the sensor. Furthermore, thicker glass substrate has greater mechanical strength, which is beneficial to restrain the stress transmission from the sensor package, and then the hysteresis error and the repeatability error can be reduced. However, there is limitions in dicing process and internal dimension of the sensor package. As a result, a trade-off has been made and a 2 mm Pyrex 7740 glass substrate has been selected.

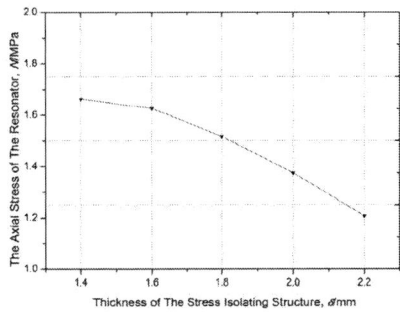

Figure 2. The axial stress of the resonator with different thickness of the stress isolating structure while the temperature changes from 20 ℃ to 60 ℃.

---

*Research supported by the National Natural Science Foundation of China (51405388, 51775447 and 51705430) and the Fundamental Research Funds for the Central Universities (G2017KY0102).

Sen Ren is with Key Laboratory of Micro/Nano Systems for Aerospace, Ministry of Education, Northwestern Polytechnical University, Xi'an, Shaanxi 710072 China (corresponding author to provide phone: +86-29-8846053; fax: +86-29-88492840; e-mail: rensen@nwpu.edu.cn).

Jianbing Xie, Qiang Shen, Fei Wang and Weizheng Yuan are with Key Laboratory of Micro/Nano Systems for Aerospace, Ministry of Education, Northwestern Polytechnical University, Xi'an, Shaanxi 710072 China (e-mail: xiejb@nwpu.edu.cn, shenq@mail.nwpu.edu.cn, wangfei@mail.nwpu.edu.cn, yuanwz@nwpu.edu.cn).

Jinkuan Zhang is with Flight Automatic Control Research Institute，Xi'an, Shaanxi 710065 China (e-mail: guangxb@facri.com).

III. FABRICATION PROCESS

The major microfabircation process steps of the atmospheric packaged SOI resonant pressure sensor is depicted in Fig. 3. The process starts with a double-side polished SOI wafer, on which a SiO_2/Si_3N_4 protective layer is deposited by using thermal oxidation and LPCVD after RCA cleaning. Then the SiO_2/Si_3N_4 protective layer is patterned to define the anisotropic etching mask. Next, the TMAH etching is carried out into the handle layer to form the diaphragm, and the SiO_2/Si_3N_4 protective layer is removed. Then, the device layer is patterned with DRIE etching. After dicing, the BOX layer is HF selectively released from which the resonator and the electrodes are formed. Finally, the SOI chip is anodic bonded onto a thick Pyrex 7740 glass substrate which has a vent hole inside, and the sensor chip is conventional atmospheric packaged. The measured pressure will be applied to the diaphragm through the pressure port of the Kovar base and the vent hole of the glass substrate. The pictures of the SOI resonant pressure sensor chip and the sensor package are shown in Fig. 4.

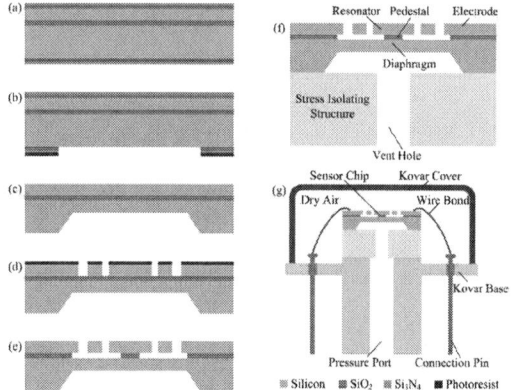

Figure 3. Microfabrication process steps of the atmospheric packaged SOI resonant pressure sensor.

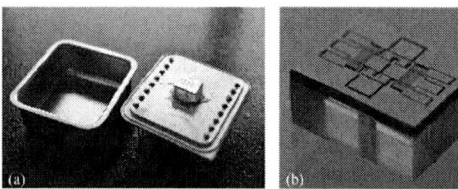

Figure 4. (a) Picture of the decaped sensor package. (b)Picture of the SOI resonant pressure sensor chip.

IV. EXPERIMENTAL RESULTS

In order to confirm the anchor design, the vertical position variations of the resonators based on two typical anchors have been measured, using Wyko NT1100 optical profiling system. The test location is on one side of the resonator by the method of testing the height difference between the stationary comb fingers and the movable comb fingers. When the pressure applied to the diaphragm changes from 2.5 kPa to 300 kPa, the vertical position of the resonator is form -1.92 μm to 3.85 μm for the conventional anchor design without suspended connecting truss, while which is from -0.24 μm to 0.28 μm for the special anchor design using suspended connecting truss. According to the experimental result, the vertical position variation is reduced to only 7.3% when the special anchor

design is used, which ensures the stability of the sensor drive and detection, and the complexity of closed-loop control is reduced.

An open-loop measurement and a closed-loop integration test have been performed for the resonant pressure sensor. Measurement dates show that the fundamental resonant frequency is from 34.5 kHz to 35.5 kHz. The variation of resonant frequency with applied pressure is shown in Fig. 5, and the performance of the atmospheric packaged SOI resonant pressure sensor has been improved by well stress isolation using thick Pyrex 7740 glass substrate. Over the full scale pressure range of 3.5–280 kPa, the pressure sensitivity is 10.86 Hz/kPa, with the frequency shift of 8.27%. The nonlinearity is 0.0138%FS, the hysteresis error is 0.0047%FS, the repeatability error is 0.0071%FS, and the accuracy is better than 0.02%FS, when a second-order polynomial fitting is used. The average temperature drift of the resonator is form −0.029%/°C to −0.012%/°C in the range of −20–60°C. The sensor has a warm-up time of 30 min, the frequency hopping is 0.2 Hz, and the sensor meets the requirement of pressure overload. Furthermore, five resonant pressure sensors have been selected randomly for uniformity test. The testing results are shown in the table I below, and the accuracy of all the resonant pressure sensors are within 0.02%FS.

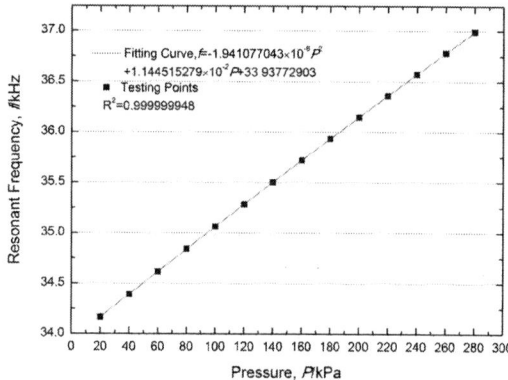

Figure 5. Resonant frequency of the atmospheric packaged SOI resonant pressure sensor with applied pressure at 20 °C.

TABLE I. TESTING RESULTS OF FIVE SOI RESONANT PRESSURE SENSORS RANDOMLY SELECTED

No.	Nonlinearity （%FS）	Hysteresis （%FS）	Repeatability （%FS）	Sensitivity （Hz/kPa）
1#	0.0138	0.0047	0.0071	10.86
2#	0.0149	0.0094	0.0128	10.94
3#	0.0112	0.0033	0.0236	11.66
4#	0.0139	0.0095	0.0182	10.96
5#	0.0135	0.0078	0.0198	9.87

REFERENCES

[1] Kinnell P K, Craddock R, "Advances in Silicon Resonant Pressure Transducers," Procedia Chemistry, vol.1, 2009, pp. 104-107.

[2] Sen Ren, Weizheng Yuan, Dayong Qiao, Jinjun Deng, Xiaodong Sun, "A micromachined pressure sensor with integrated resonator operating at atmospheric pressure," Sensors, vol. 4, Dec. 2013, pp. 17006-17024.

[3] Sen Ren, Weizheng Yuan, Xiaodong Sun, Jinjun Deng, Dayong Qiao, Chengyu Jiang, "Microfabricated SOI pressure sensor using dynamically balanced lateral resonator," in IEEE-NEMS 2014, Hawaii, USA, 2014, pp. 229-232.

Flexible pressure and force sensing system for wearable medical devices

Ning Xue, *Member, IEEE*, Chunxiu Liu, Jianhai Sun, Chao Wang, Senior *Member, IEEE*

Abstract—**Flexible materials as substrate and/or sensing elements are required for comfortable and conformal wear of such devices. This paper reviews those fabrication technologies and sensor interface circuits for device applications on human skin. Authors' current research overview on flexible pressure sensor on medical application is also discussed.**

I. INTRODUCTION

Driving by the advance in flexible and stretchable organic materials, soft and flexible sensors have emerged in medical applications. The mismatch in mechanical properties of those medical devices on human body is becoming small. The technologies utilize both conventional microeletric-mechanical system (MEMS) and non-conventional fabrication techniques, such as soft film laminating, transfer printing, and screen printing to pattern sensors on flexible substrate suitable for the required medical applications. Noninvasive flexible pressure or force sensors are mostly commonly used for human diseases monitoring and diagnosis, such as skin patch for continuous monitoring of pulse waveform, cricoid pressure, foot pressure, bone stress. The sensing mechanism are usually based on piezoresistive sensing, capacitive sensing, and piezoelectric sensing.

This paper reviews the use of those sensing principles in diagnostic devices that mount on the skins. Our recent researches are highlighted subsequently. The challenge of the wearable and conformal sensing system are provided with some perspectives on future opportunities.

II. REVIEW OF THE CURRENT FLEXIBLE WEARABLE DEVICES

The basic principle of the pressure/force sensors are based on piezoresisitve, capacitive and piezoelectric effect. Thanks to the "single layer" of the sensor device and simplicity of its readout circuit, the piezoresisitve sensor comprised of nanocomposites or soft metals as sensing elements have attracted much attention. Typically, the nanocomposite materials include Ag, Au nano-rod, carbon black and carbon nanotube (CNT) and graphene. Figure 1 is an example of the CNT-Ag nanoparticle composite. The pressure sensitivity of 8%/Pa was reported with up to 1000 bending circles [1]. Besides, the metal based strain gauge is another type of piezoresistive device. For this, the metal pattern is usually formed by either direct photo-lithography and etching or pattern transfer technology. In the latter process, 3D buckled

N. Xue, C. Liu, J. Sun are with the Institute of Electronics, Chinese Academy Of Sciences, Beijing, China 100190 (email: xuening@mailie.ac.cn).

C. Wang are with the Singapore University of technology and Design, Singapore, 487372 (email: chao.wang.1978@ieee.org)

Au film are transferred to the host substrate. This was done by pre-stretch of the elastomeric substrate during the transfer printing. The buckled "wavy" patterned sensing elements enables over 100% strain before fracture [2] (figure 2).

Figure 1. CNT-AgNP composite film pressure sensor.(A) schematic of PDMS substrate nanoparticle pressure sensor. (B) SEM image of the composite film. (C) Stress change of the sensor with different AgNP concentrations. (D) Device stability testing before and after 500 cycles.

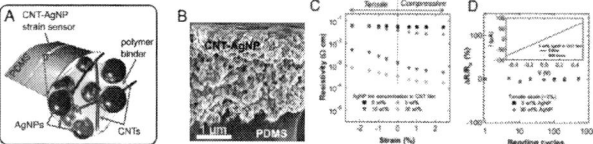

Figure 2. Schematic of the pattern transfer process. The sensing elements were released from a source wafer to a target substrate through Von der Waals force. (b) The wavy patterns. (c) The 3D structure after pattern transfer.

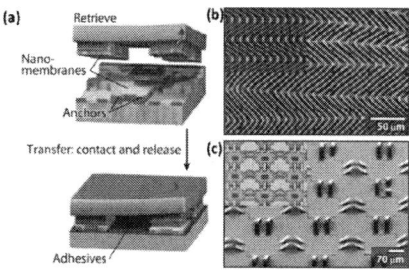

Capacitive sensing is a commonly used pressure measurement method with advantages of high sensitivity, large dynamic range and temperature independence. But the stray capacitance may cause the system susceptible to noise. Thus the electromagnetic shielding was often adopted to isolate the device from the environment. Metal-dielectric-metal sandwich capacitor are the basic sensing structure. During the fabrication, the dielectric layer can be patterned into square, pyramid structure to enhance the device sensitivity. The top metal may be formed by ether metal deposition or bonding of the two flexible substrates. Figure 3(a-b) describes the fabrication process flow and image of the capacitive based pressure sensor. Furthermore, the pressure sensitive organic transistor is used for amplification of the output signal, thus the resolution is enhanced (figure 5(c))[3].

Piezoelectric sensing is a widely used nowadays on dynamic pressure sensing, especially on microphone, gyroscope. It has benefit of power self-generation and insensitivity to the electromagnetic environment. The dynamic pressure or force such as pulse waveform monitoring can be tackled by such a device. Figure 4 demonstrates structure and application of the wearable dynamic pressure sensor based on PZT piezoelectric materials parallelly connected to an organic MOSFET output signal amplifier [4].

Figure 3. (a) Fabrication process flow of the flexible capacitive pressure sensor;(b) image of the sensor; (c) organic transistor pressure sensor.

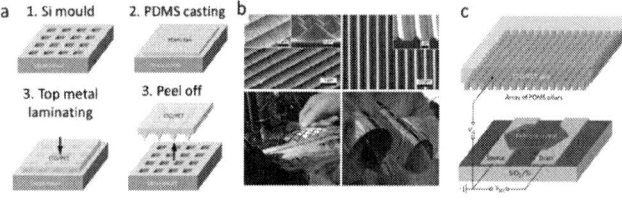

Figure 4. (a-c) Structure of the flexible PZT piezoelectric waveform monitoring device, (d-e) device conformally mounts on human skin surface.

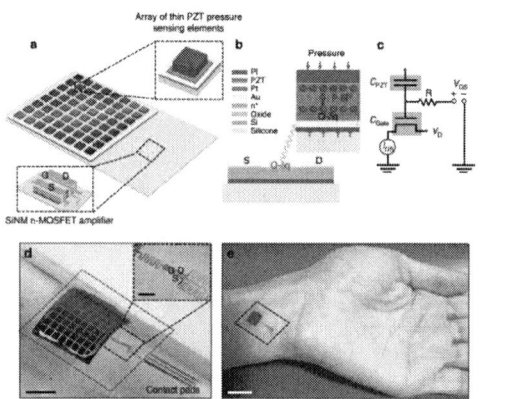

III. OUR RESEARCH HIGHLIGHT

Several human-skin friendly pressure/force sensing system aiming to solve medical issues during surgical operation and prognostic physiological monitoring are to be exhibited. The sensors are able to attach and mount on the skin conformally and tightly. Figure 5 is a cricoid force monitoring system consisting of a disposable, flexible, wearable and biocompatible PDMS-based sensor patch, flexible printed circuit board (FPCB), circuit board with microprocessor with LED indicator to display the applied force range. The sensors are capacitive based with PDMS square as pressure sensitive membrane. Cr/Au as top metal was deposited on the surface of the PDMS membrane. The bottom metal was from commercial FPCB. The force detecting circuit consists of column and row capacitor-voltage convertor, multiplexers, analog-to-digital convertor (ADC), microprocessor and LED force display. The 100 channel pressure signal was acquired and processed in microprocessor to calculate the corresponding applied force. The system has the resolution of 1N within 0-50 N range. The results are capable to display in computer or LED every 0.5 sec.

We also developed an AlN based flexible piezoelectric pressure for real-time pulse waveform monitoring. The AlN based elements was fabricated on the PDMS substrate. As pulse wave applied on the sensor, the electrical charges on the sensor is generated. Figure 6 is the schematic view of the flexible piezoelectric pressure sensor. The detection circuits consist of signal conditioning circuit with a two-stage

operational amplifier. The sensor was currently under fabrication and the circuit was tested with capability of $0.7 \mu V_{rms}$ resolution, 0.5-100Hz bandwidth, and the circuit noise is lower than $0.7 \mu V_{rms}$.

Figure 5. Cricoid pressure/force sensing system with image of capacitive sensor, (b) processing and display circuit, (c) calibration and testing on manikin neck skin surface.

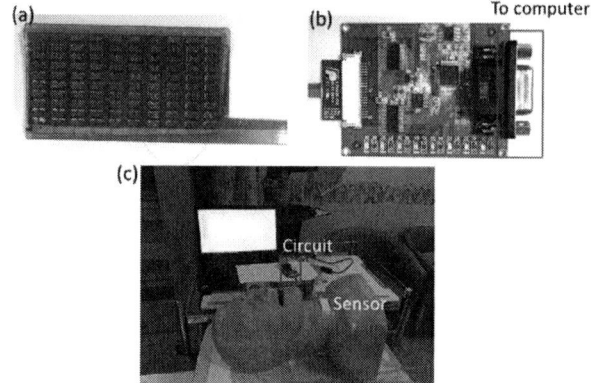

Figure 6. (a) Schematics of the AlN based piezoelectric pressure sensor, (b) the conditional circuit for signal modification and amplification.

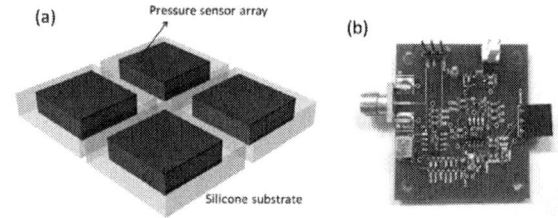

IV. CONCLUSION

The state-of-art skin-friendly flexible pressure/force sensors were summarized. Three types (capacitance, piezoresistive, and piezoelectric) of pressure sensors were developed for particular medical applications. The measurement results of the system were discussed. Currently, the pressure sensors on the flexible substrate has been promising to solve many medical issues. The sensor integrated with the flexible processing circuit is the next forwarding step. We believe that our methodologies and efforts would inspire the development of more future wearable pressure sensors and even wearable, soft electronics to serve for biomedical application.

REFERENCES

[1] K. Takei, Z. Yu, M. Zheng, et al, PNAS, vol. 111, no. 5, pp. 1703-1707, 2014
[2] W.M. Choi, J. Song, D.Y. Khang, et al, *Nano Lett.* Vol. 7, pp.1655–63, 2007.
[3] S. C. B. Mannsfeld, B. C.-K. Tee, R. M. Stoltenberge, et al, nature materials, vol. 9, pp. 859-864, 2010.
[4] C, Dagdeviren, Y. Su, P. Joe, et al, nature communication, vol. 5, pp. 4496, 2014.

Preparation of multi-walled carbon nanotubes/polydimethylsiloxane composite for electronic skin application

Cheng Chi, Xuguang Sun, Tong Li, Ning Xue * and Chang Liu *

Abstract— We compared the six common organic solvents in the solution blending method to prepare MWCNT/PDMS composite, including toluene, ethanol, chloroform, n-hexane, tetrahydrofuran (THF) and dimethylformamide (DMF). We proposed the best choice of organic solvents by comparing the dispersion performance of MWCNT and the stability of MWCNT/PDMS. Then, we present the preparation and micromolding of MWCNT/PDMS composite.

Index Terms— Electronic skin, tactile sensing, nanocomposite, carbon nanotube.

I. INTRODUCTION

Because of excellent electrical, thermal and mechanical properties, Carbon nanotubes (CNTs) have been widely used in the fields of materials, biomedicine, and micro-nano devices since the discovery of CNTs in 1991 by Iijima [1]. The incorporation of CNTs into polymers to prepare nanocomposites has received extensive attention in recent years. This kind of nanocomposites has a good piezoresistive effect under the premise of preserving the flexibility and extensibility of the polymer, and the related research has become an important research direction of electronic skin. Compared to other nanofillers, such as nano-silver particles [2] and carbon black (CB) [3], CNTs have an ultra-high aspect ratio (typically above 1000: 1), making the CNTs/polymer material could obtain a good conductive capacity at a lower doping amount, which greatly improves the mechanical properties of composite materials. Polydimethylsiloxane (PDMS) is an inert, non-toxic polymer organosilicon compound. Because of its good elasticity and biocompatibility, PDMS is widely used in the micro-channel structure design of micro-electromechanical systems (MEMS) chips.

The biggest challenge in the preparation of CNTs/PDMS composites is to overcome the phenomenon of aggregation and to distribute CNTs evenly in PDMS against van der Waals forces in CNTs. In recent years, the common preparation methods include melt blending method [4], polymerization method [5] and solution blending method. Among them, the solution blending method has a good application prospect

---

*Research supported by the Recruitment Program of Global Experts and Frontier Science Key Program of the Chinese Academy of Sciences (QYZDY-SSW-JSC037).

Cheng Chi, Xuguang Sun and Ning Xue are with the State Key Laboratory of Transducer Technology, Institute of Electronics Chinese Academy of Sciences, Beijing 100190, China, and also with the School of Electronic, Electrical, and Communication Engineering, University of Chinese Academy of Sciences, Beijing 100190, China (e-mail: chicheng15@mails.ucas.ac.cn; sunxuguang16@mails.ucas.ac.cn; xuening@mail.ie.ac.cn).

Tong Li and Chang Liu are with the State Key Laboratory of Transducer Technology, Institute of Electronics Chinese Academy of Sciences, Beijing 100190, China (phone: +86-10-5888-7638; e-mail: tli@mail.ie.ac.cn; changliu8888@gmail.com).

---

because of its simple operation and good dispersion effect. The key of this method is the choice of organic solvent, which is necessary to make CNTs have good dispersion effect in them, and dissolve PDMS better. In recent years, the organic solvents used in the literature consist of toluene [6], ethanol , chloroform [7], n-hexane [8], tetrahydrofuran (THF) [9]and dimethylformamide (DMF) [10].

In this paper, we compared the dispersion performance of MWCNT and the stability of MWCNT/PDMS in the six organic solvents. The CNTs/PDMS composite was fabricated on a printed circuit board with gold electrode with a thickness of 20μm using micromachining.

II. EXPERIMENTS AND RESULTS

The MWCNT in toluene and n-hexane appeared precipitation soon after sonication, indicating that the dispersion of MWCNT in these two solutions was poor and these two solvents are not suitable to be the common solvents. However, chloroform, DMF, THF and ethanol had a good dispersion effect on MWCNT, their solution could keep steady for longer than two weeks. Since DMF reacts with PDMS to produce white colloidal material, and PDMS is insoluble in ethanol, DMF and ethanol are not suitable either. Then, we compared the stability of MWCNT/PDMS in chloroform and THF. Chloroform performed much better than THF. Consequently, chloroform is the best choice in the six test solvents.

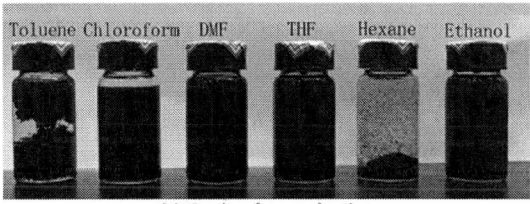

(a) 5 min after sonication

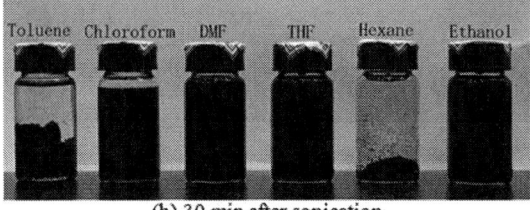

(b) 30 min after sonication

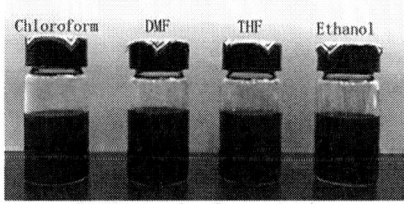

(c) Two weeks after sonication

Fig. 1 The dispersion effect of MWCNT dispersed in different solvents after a period of time

The CNTs / PDMS composites were fabricated on a PCB circuit board with gold electrode with a thickness of 20µm. The illustration of the fabrication process is shown in Fig. 2.

We used silicon, PDMS and PS/PTFE mold to transfer structure successively. This method could reduce the viscosity through the process and achieve high-quality transfer. The SEM image of the cross section of MWCNT/PDMS is shown in Fig. 3. MWCNTs are dispersed evenly in PDMS.

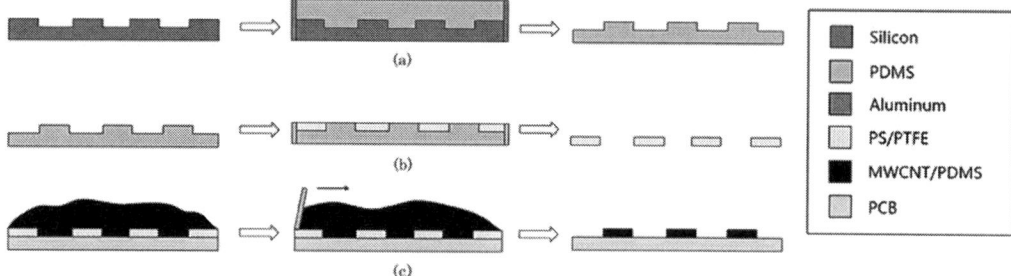

Fig. 2 Micromachining fabrication process of MWCNT/PDMS composite

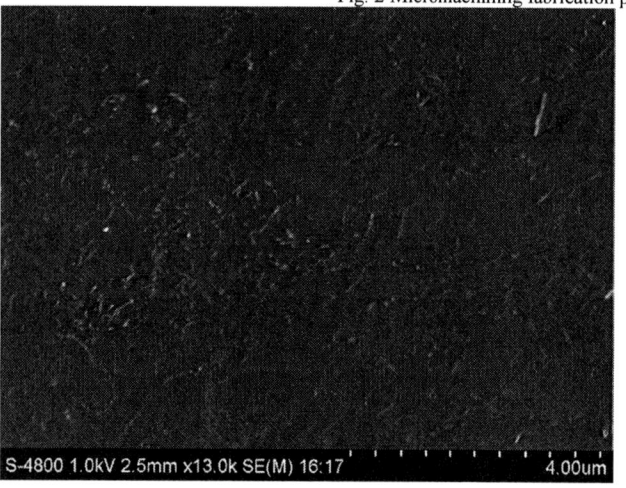

S-4800 1.0kV 2.5mm x13.0k SE(M) 16:17   4.00um

Fig. 3 SEM image of the cross section of MWCNT/PDMS

III. CONCLUSION

In this work, we compared the six common organic solvents in the solution blending method to prepare MWCNT/PDMS composite. And we demonstrated that chloroform was the best choice of organic solvents by comparing the dispersion performance of MWCNT and the stability of MWCNT/PDMS. Then, we patterned the MWCNT/PDMS composite on printed circuit board using micromolding method.

We just realized a simple rectangular shape of the structure transfer. In future work, effective and robust structure is necessary to achieve tactile sensing over the whole body. Meanwhile, scanning circuit and tactile signal processing algorithm are also required.

[1]   S. Iijima, "Helical microtubules of graphitic carbon," *Nature*, vol. 354, pp. 56–58, 1991.

[2]   X. Niu, S. Peng, L. Liu, W. Wen, and P. Sheng, "Characterizing and patterning of PDMS-based conducting composites," *Adv. Mater.*, vol. 19, no. 18, pp. 2682–2686, 2007.

[3]   J. J. Wang, M. Y. Lin, H. Y. Liang, and R. Chen, "Piezoresistive nanocomposite rubber elastomer for stretchable MEMS sensor," in *IEEE International Conference on MICRO Electro Mechanical Systems*, 2016, pp. 550–553.

[4]   R. Andrews, D. Jacques, M. Minot, and T. Rantell, "Fabrication of Carbon Multiwall Nanotube/Polymer Composites by Shear Mixing," *Macromol. Mater. Eng.*, vol. 287, no. 6, pp. 395–403, 2015.

[5]   R. Allen, L. Pan, G. G. Fuller, and Z. Bao, "Using in-situ polymerization of conductive polymers to enhance the electrical properties of solution-processed carbon nanotube films and fibers," *ACS Appl. Mater. Interfaces*, vol. 6, no. 13, pp. 9966–9974, 2014.

[6]   L. Wang *et al.*, "PDMS/MWCNT-based tactile sensor array with coplanar electrodes for crosstalk suppression," *Microsystems Nanoeng.*, vol. 2, p. 16065, 2016.

[7]   J. Hwang *et al.*, "Poly (3-hexylthiophene) wrapped carbon nanotube/poly (dimethylsiloxane) composites for use in finger-sensing piezoresistive pressure sensors," *Carbon N. Y.*, vol. 49, no. 1, pp. 106–110, 2011.

[8]   L. Wang, X. Wang, and Y. Li, "Relation between repeated uniaxial compressive pressure and electrical resistance of carbon nanotube filled silicone rubber composite," *Compos. Part A Appl. Sci. Manuf.*, vol. 43, no. 2, pp. 268–274, 2012.

[9]   J. Hong, J. Lee, C. K. Hong, and S. E. Shim, "Effect of dispersion state of carbon nanotube on the thermal conductivity of poly(dimethyl siloxane) composites," *Curr. Appl. Phys.*, vol. 10, no. 1, pp. 359–363, 2010.

[10]   U. Subramanyam and J. P. Kennedy, "PVA networks grafted with PDMS branches," *J. Polym. Sci. Part A Polym. Chem.*, vol. 47, no. 20, pp. 5272–5277, 2009.

Water Hardness Determination Using Disposable MEMS-Based Electrochemical Sensor

Nan Wang, Elgar Kanhere, Kai Tao, Jin Wu, Jianmin Miao, and Michael S. Triantafyllou

Abstract—**This paper presents a compact and disposable electrochemical sensor which can be batch fabricated by standard microfabrication technology. The proposed sensor has the potential to be directly deployed for measuring water hardness, which is mainly contributed by dissolved calcium (Ca) and magnesium (Mg) ions. The analytical performance of the sensor is evaluated through electrochemical experiments. The experimental results indicate that the sensor is capable of detecting Ca and Mg ions down to 1 ppm with a linear detection range from 10 to 50 ppm.**

I. INTRODUCTION

In high-rise buildings, maintaining thermal comfort for people who live or work inside the buildings is one of the major concerns. Water-based cooling towers have been widely incorporated into high-rise buildings to remove unwanted heat from the surroundings. One critical challenge for the cooling tower is the formation of scale in the pipeline system. The scale is gradually aggregated when dissolved solids crystallize and precipitate on the surface of valves and tubes. Among all the dissolved solids, calcium (Ca) and magnesium (Mg) minerals, which are the main contributors to water hardness, play the most significant role in the scale formation. As the thickness of such tenacious scale increases over time, it will adversely affect the flow distribution, resulting in a drastic decrease of heat transfer capability. Therefore, the development of miniaturized sensors, which can be directly used to monitor the concentration of dissolved Ca and Mg ions, are of great interest to industrial researchers.

In recent years, electrochemical sensors/devices based on stripping technique have become promising measuring tools in the assessment of environmental pollutants [1-3]. The remarkable sensitivity of stripping technique arises from the productive deposition step, which ensures sufficient amount of analyte of interest to be accumulated on the sensing electrode.

Nan Wang is with the Center for Environmental Sensing and Modeling IRG, Singapore-MIT Alliance for Research and Technology Centre, 138602 Singapore (phone: 65-6516-6129; fax: 65-6684-2118; e-mail: WANG0845@e.ntu.edu.sg).

Elgar Kanhere is with the School of Mechanical and Aerospace Engineering, Nanyang Technological University, 639798 Singapore (e-mail: ELGARVIK001@e.ntu.edu.sg).

Kai Tao is with the Department of Microsystem Engineering, Northwestern Polytechnical University, 710072 China (e-mail: taokai@nwpu.edu.cn).

Jin Wu is with the School of Electronics and Information Technology, Sun Yat-sen University, 510275 China (e-mail: jwu6@ntu.edu.sg).

Jianmin Miao is with the School of Mechanical and Aerospace Engineering, Nanyang Technological University, 639798 Singapore (e-mail: jmiao@pmail.ntu.edu.sg).

Michael S. Triantafyllou is with the Department of Mechanical Engineering, Massachusetts Institute of Technology, Cambridge, MA 02139 USA (e-mail: mistetri@mit.edu).

Another merit is that electrochemical sensors can be batch fabricated using classical microelectromechanical systems (MEMS) manufacturing processes [4-5], such as photolithographic patterning, thin film deposition, dry and wet etching, etc. In this paper, a compact and disposable electrochemical sensor is proposed to be potentially used for water hardness determination. The sensor is constructed by standard MEMS technology. The analytical performance of the sensor to detect Ca and Mg ions in aqueous solution is explored with the aid of electrochemical testing.

II. SENSOR FABRICATION

The disposable MEMS-based electrochemical sensors were fabricated on top of a silicon wafer. An insulation silicon dioxide layer was initially deposited by the plasma-enhanced chemical vapor deposition method. The wafer was then thoroughly cleaned by acetone and rinsed with deionized (DI) water. The washed wafer was dried in a rotational spin dryer, after which the wafer was transferred into a hexamethyldisilazane promoter. After cooling, a layer of photoresist was spin-coated on the surface of the wafer followed by baking on a hotplate for 4 min. The baked photoresist was exposed to i-line ultraviolet light. After exposure, the wafer was immersed in developer solution to remove the exposed photoresist, during which agitation was provided to promote the developing. After rinsing with DI water and drying by nitrogen (N_2) gas, the wafer was placed in the chamber of a magnetron sputtering machine to deposit metal layers on top of the wafer, forming reference, working, and counter electrodes for the sensor. After that, the wafer was completely dipped in acetone for 12 hrs. All remaining photoresist left on the wafer and the sputtered metal attached on the photoresist were totally removed by the acetone. Subsequently, the wafer was again rinsed with DI water and dried by N_2 gas to wash away all residues. Finally, the wafer was diced and the individual sensor was packaged.

III. EXPERIMENTS AND RESULTS

To investigate the analytical performance of the disposable MEMS-based electrochemical sensor for water hardness detection, a series of square wave anodic stripping voltammetry (SWASV) experiments were conducted. The experimental setup is shown in Figure 1, in which the packaged sensor was immersed in aqueous solution containing equal concentrations of Ca and Mg ions. The wires of reference, working, and counter electrodes of the sensor were connected to a CHI 600C electrochemical workstation using crocodile clips. All experiments were controlled using the CH Instruments software installed on a computer, through which different experimental parameters were specified.

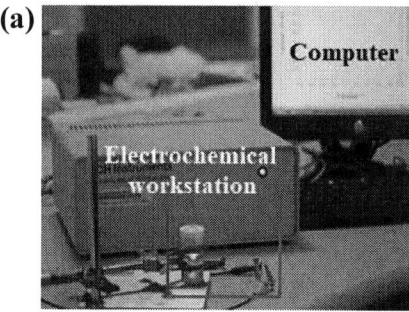

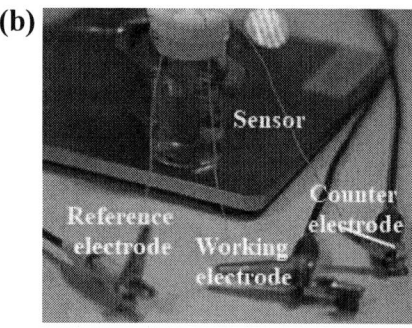

Figure 1. (a) Photograph of the experimental setup to investigate the analytical performance of the disposable MEMS-based electrochemical sensor. (b) An enlarged view of the red box highlighted in (a) to show the configuration of the SWASV experiments.

The parameters for the SWASV experiments were selected as deposition potential of -0.5 V, deposition time of 120 s, frequency of 50 Hz, amplitude of 25 mV, and step potential of 5 mV. The responses of the sensor were recorded from -0.4 to 0.9 V by increasing the concentrations of both Ca and Mg ions from 10 to 50 ppm. As illustrated in Figure 2, legible stripping peaks (near the potential of 0.2 V) can be observed from all the voltammograms. The magnitude of stripping peak currents with respect to the concentrations of Ca and Mg ions is plotted in Figure 3, where a linear correlation (R^2=0.97) is obtained. The minimum distinguishable stripping peak for the sensor was recorded when the concentration of Ca and Mg ions was 1 ppm.

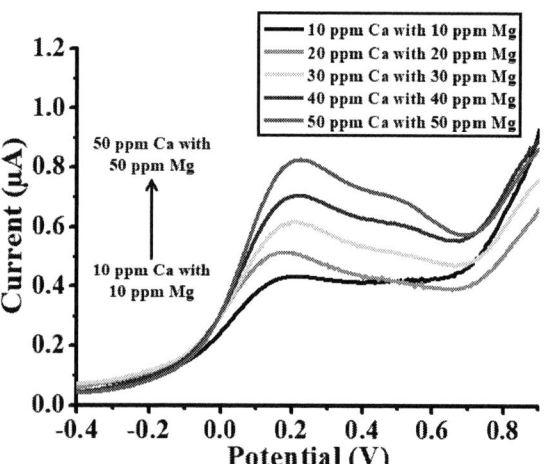

Figure 2. Anodic stripping voltammograms recorded for the disposable MEMS-based electrochemical sensor with increased concentrations of Ca and Mg ions from 10 to 50 ppm.

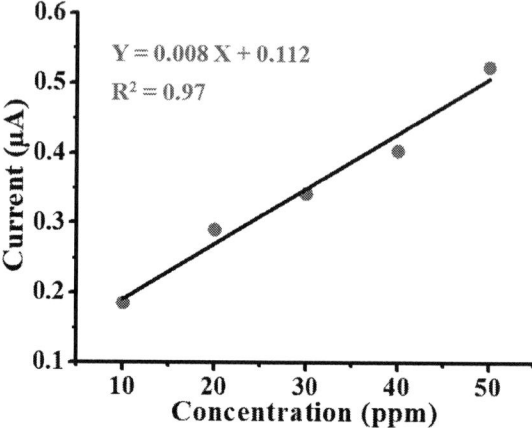

Figure 3. Calibration plot of the magnitude of stripping peak currents with respect to the concentrations of Ca and Mg ions.

IV. CONCLUSION

In this paper, a disposable electrochemical sensor was proposed and fabricated by means of standard MEMS manufacturing processes. The capability of the sensor to determine Ca and Mg ions in aqueous solution was demonstrated by electrochemical experiments. The sensor exhibits favorable detection limit of 1 ppm toward Ca and Mg ions as well as linear response within the concentration range from 10 to 50 ppm. The experimental results manifest a promising application of the sensor for water hardness measurement.

ACKNOWLEDGMENT

This research is supported by the National Research Foundation (NRF), Prime Minister's Office, Singapore under its Campus for Research Excellence and Technological Enterprise (CREATE) programme. The Center for Environmental Sensing and Modeling (CENSAM) is an interdisciplinary research group (IRG) of the Singapore MIT Alliance for Research and Technology (SMART) centre.

REFERENCES

[1] N. Wang, E. Kanhere, M. S. Triantafyllou, and J. M. Miao, "Shark-inspired MEMS chemical sensor with epithelium-like micropillar electrode array for lead detection," in *18th International Conference on Solid-State Sensors, Actuators and Microsystems*, Anchorage, 2015, pp. 1464-1467.

[2] N. Wang, E. Kanhere, M. S. Triantafyllou, and J. M. Miao, "Copper detection with bio-inspired MEMS-based electrochemical sensor," in *19th International Conference on Miniaturized Systems for Chemistry and Life Sciences*, Gyeongju, 2015, pp. 23–25.

[3] N. Wang, M. Kitajima, K. Mani, E. Kanhere, A. J. Whittle, M. S. Triantafyllou, and J. M. Miao, "Miniaturized electrochemical sensor modified with aptamers for rapid norovirus detection," in *11th IEEE Annual International Conference on Nano/Micro Engineered and Molecular Systems*, Sendai, 2016, pp. 587-590.

[4] E. Kanhere, N. Wang, M. Asadnia, A. G. P. Kottapalli, and J. M. Miao, "Crocodile inspired dome pressure sensor for hydrodynamic sensing," in *18th International Conference on Solid-State Sensors, Actuators and Microsystems*, Anchorage, 2015, pp. 1199–1202.

[5] K. Tao, S. W. Lye, N. Wang, X. Hu, and J. M. Miao, "A sandwich-structured MEMS electret power generator for multi-directional vibration energy harvesting," in *18th International Conference on Solid-State Sensors, Actuators and Microsystems*, Anchorage, 2015, pp. 51–54.

The Fabrication Method of High Height-to-Diameter Ratio Micro-Wineglass Resonators*

Jianbing Xie, *Member, IEEE*, Sen Ren, Qiang Shen, Lei Chen, Hui Xie, Weizheng Yuan

Abstract—This paper presents the fabrication of High Height-to-Diameter Ratio Micro-Wineglass Resonators using glassblowing with thermal decomposition process. *CaCO₃* is used as the thermal decomposition material in this process in this paper. A number of chips with 200μm deep cavities are fabricated and loaded with different doses of *CaCO₃*. The experiment results show that, the 200μm deep cavity can blowing a glass spherical shell with the H/DR of 0.79, which is better than the 800μm deep cavity glassblowing process without thermal decomposition.

Keywords—Micro-Wineglass Resonators, Hemispherical Resonator Gyroscopes, Glassblowing, Height-to-Diameter Ratio, Thermal Decomposition

I. INTRODUCTION

Micro Hemispherical Resonator Gyroscopes (μHRG) is a new type of inertial sensor with high precision potential. Micro-scale Glassblowing has proved to be effective to fabricate mushroom and bubble shaped spherical structures [1-5]. However, the fabrication of high smoothness, high-accuracy roundness and high height-to-diameter ratio (H/DR) 3-D micro-wineglass resonators remains to be a challenge. In order to obtain high H/DR 3-D micro-wineglass resonator, several fabrication method have been reported in the literature. In [1, 2, 3], cylindrical cavities are used to blowing glass shells, as shown in Fig.1 (a). In order to make the shells as spherical as possible, it is necessary to use thick wafers and etch deep cavities. In [1, 4, 5], alternative fabrication process utilizing two bonded silicon wafers is used to achieve larger high H/DR, as shown in Fig.1 (b). In this process, a thin wafer with a small through hole is used to define the opening of the glassblowing, and a thick wafer with a large chamber is used to provide the gas required of blowing. Although the above methods can get good H/DR, but the processes are complex and expensive.

The purpose of this paper is to demonstrate the feasibility of the fabrication of high H/DR 3-D micro-wineglass resonators using thermal decomposition glassblowing process. As shown in Fig. 1 (c). The thermal decomposition materials are sealed in tie shallow cavities before

---

*Resrach supported by National Natural Science Foundation of China (51405388, 51775447, 51705430), and Fundamental Research Funds for the Central Universities (G2017KY0102).

The authors are with the Key Laboratory of Micro/Nano Systems for Aerospace, Ministry of Education, Northwestern Polytechnical University P.O.Box 638, 127# Youyi West Road, Xi'an 710072, Shaanxi, China. (e-mail: xiejb@nwpu.edu.cn; rensen@nwpu.edu.cn; shenq@nwpu.edu.cn; cl@mail.nwpu.edu.cn; 547567184@qq.com; yuanwz@nwpu.edu.cn).

---

glassblowing, when the glassblowing process begins, the materials will be decomposed to blow a big bubble with high H/DR. The advantage of this method is that the cost of the process is reduced, and easy to get high H/DR 3-D micro-wineglass resonators.

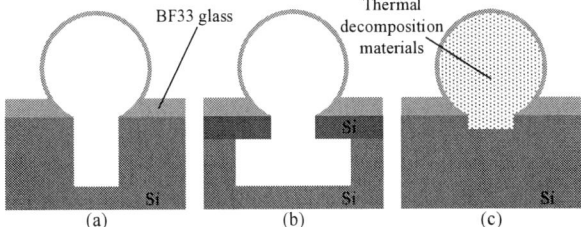

Figure 1. 3-D micro-scale glassblowing process, (a)with silicon deep etching only, (b)with silicon deep etching and bonding (c) with silicon shallow etching and thermal decomposition

II. PROCESS PRINCIPLE

In this paper, the glassblowing process consists of the following: 1) the thermal decomposition material are deposit into the pre-etched cavities of silicon wafer; 2) anodic bonding of the BF33 glass wafer and silicon wafer; 3) heating the wafer stack above the softening point of glass to blowing the micro-wineglass shells.

CaCO₃ is used as the thermal decomposition material in this process, because the thermal decomposition temperature of which is lower than blowing and higher than anodic bonding. The thermal decomposition reaction equation is described as follows:

$$CaCO_3(s) \rightarrow CaO(s) + CO_2\uparrow$$

The solid *CaCO₃* cannot deposited into the cavities directly, so the precipitation reaction of *NaCO₃* solution and *CaCl₂* solution is used, and the chemical equation is:

$$Na_2CO_3(aq) + CaCl_2(aq) \rightarrow CaCO_3\downarrow + Na_2Cl_2(aq)$$

After the precipitation reaction, the solid CaCO3 are deposited into the cavities, and then through the subsequent drying and anodic bonding, the resonator wafer before blowing is shown in Fig.2.

$CaCO_3$

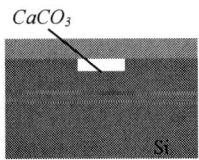

Figure 2. The bonded resonator chip before blowing

III. FABRICATION OF HIGH H/DR MICRO-WINEGLASS RESONATORS

In this paper, a 500μm double polished single crystal silicon wafer and a 100μm BF33 glass wafer are used to fabricate the high H/DR micro-wineglass resonators.

Firstly, 200μm deep cavities are etched on the silicon wafer using DRIE. Then, different volumes of $NaCO_3$ solution and $CaCl_2$ solution are injected into the cavities through an "nL" injection pump as shown in Fig.3.

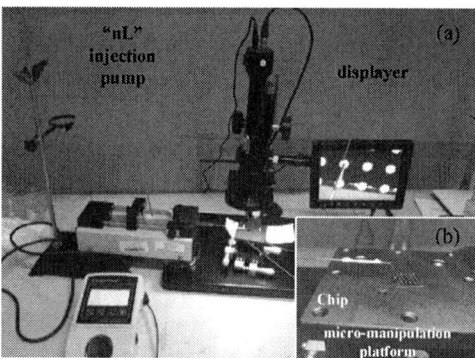

Figure 3. (a)"nL" injection system and (b)close-up of injection chip

In order to verify the effect of $CaCO_3$ dose on the H/DR, we designed a number of experiments with different doses of $CaCO_3$. The experimental results are shown in the Fig.4 and Tab.1. From *Sample 1* to *Sample 4*, the height of glass-blown spherical shell increased with the increase of $CaCO_3$ weight, and the H/DR increased from 0.12 to 0.79, which illustrate that, the using thermal decomposition glassblowing process can effectively increase the H/DR of glass-blown. For *Sample 5*, the blown spherical shells are crowded together and crushed due to excess blowing.

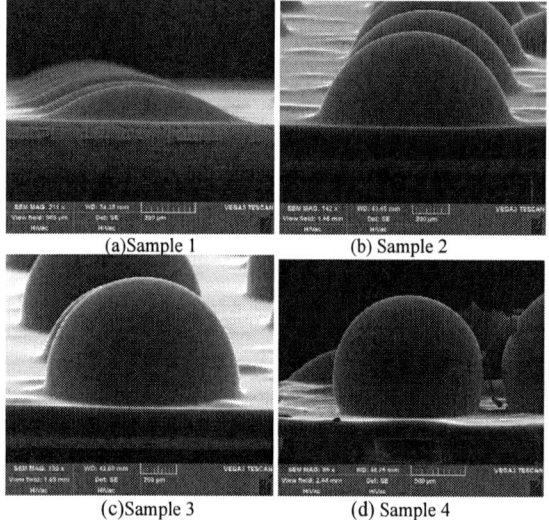

(a)Sample 1   (b) Sample 2

(c)Sample 3   (d) Sample 4

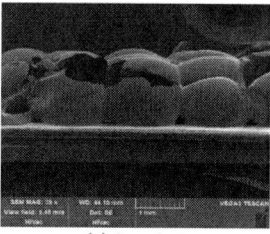

(e) Sample 5

Figure 4. SEM images of glass-blown spherical shell

TABLE I. EXPERIMENTA DATA

	m_{CaCO_3} (μg)	H (μm)	D (μm)	H/DR
Sample 1	none	133.39	1088.34	0.12
Sample 2	0.236	516.26	1026.86	0.50
Sample 3	0.472	682.42	1065.13	0.64
Sample 4	1.887	1107.82	1397.51	0.79
Sample 5	2.830	/	/	/

IV. CHARACTERIZATION

As a comparison, Fig.5 shows the fabricated H/DR 3-D micro-wineglass resonators with (a) and without (b) thermal decomposition glassblowing process.

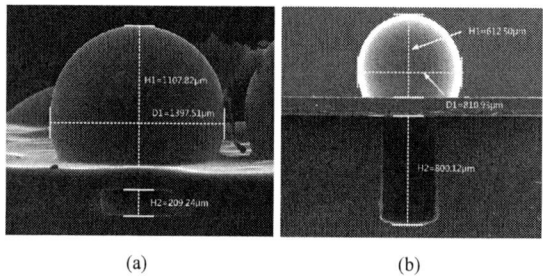

(a)   (b)

Figure 5. Fabricated H/DR 3-D micro-wineglass resonators with(a) and without(b) thermal decomposition glassblowing process Conclusion

The experiment results show that, using the $CaCO_3$ thermal decomposition glassblowing process, the 200μm deep cavity can blowing a glass spherical shell with the H/DR of 0.79, which is better than the 800μm deep cavity glassblowing process without thermal decomposition. Furthermore, an H/DR more than 0.9 can be achieved through adding proper thermal decomposition materials.

REFERENCES

[1] Eklund E. J, Shkel A. M. Glass blowing on a wafer level [J]. Journal of Microelectromechanical Systems, 2007, 16(2): 232-239.

[2] Giner J, Gray J. M, Gertsch J, et al. Design, fabrication, and characterization of a micromachined glass-blown spherical resonator with insitu integrated silicon electrodes and ALD tungsten interior coating[C]//Micro Electro Mechanical Systems (MEMS), 2015 28th IEEE International Conference on. IEEE, 2015: 805-808.

[3] Senkal D, Ahamed M. J, Trusov A. A, et al. Electrostatic and mechanical characterization of 3-D micro-wineglass resonators[J]. Sensors and Actuators A: Physical, 2014, 215: 150-154.

[4] Prikhodko I. P, Zotov S. A, Trusov A. A, et al. Microscale glass-blown three-dimensional spherical shell resonators [J]. Journal of Microelectromechanical Systems, 2011, 20(3): 691-701.

[5] Senkal D, Prikhodko I. P, Trusov A. A, et al. Micromachined 3-D glass-blown wineglass structures for vibratory MEMS applications [J]. Technologies or Future Micro-Nano Manufacturing, 2011: 8-10.

In-plane Rotational Tuning of Polymer Diffraction Grating for Diverse Imaging Spectroscopy

Sanathanan S. Muttikulangara[1], Maciej Baranski[1], Shakil Rehman[2], Liangxing Hu[1] and Jianmin Miao[1]

Abstract—**This paper discusses fabrication and testing of a monolithic rotary actuator that works on the principle of electrostatic stepper motor mechanism. Optical diffraction gratings are fabricated using negative SU-8 photoresist which are to be embedded on the rotary actuator for tuning.**

I. INTRODUCTION

Spectroscopy is a means of studying the properties of physical objects based on measuring how an object emits and interacts with visible, ultraviolet, or infrared light. It could also be an optical remote sensing technique involving the acquisition of information such as chemical composition and temperature of an object without coming into physical contact with that object. Optical spectroscopy is therefore widely used in imaging satellites, as the spectroscopic information identifies chemical composition, temperature and other properties of the atmosphere, land, and water beneath.

Imaging spectrometry is used in remote sensing for determining the chemical composition of the imaged objects. The three-dimensional information (2-spatial and 1-spectral) are measured by techniques like, point (whisk broom) scanning, line (push broom) scanning, spectral scanning or snapshot imaging. One of the methods of achieving scanning is by tuning the diffraction grating which is an integral component in the spectrometer. Microelectromechanical Systems (MEMS) fabrication technology is a promising approach for making miniaturized devices [1]. Previously, there were attempts in making tunable diffraction grating using micromachining technology. Tuning can be achieved by either changing the pitch [2] or by out of plane scanning (tilt scanning). Most of the tunable gratings in MEMS are realized with silicon-based technology are having relatively low diffraction efficiency and the period of the diffraction grating is restricted to photolithographic resolution. Further, it is challenging to fabricate blazed angles diffraction grating using silicon micromachining. Another challenge in MEMS-based tuning devices is their high operational frequency (few kHz) which makes it difficult to employ them with conventional camera, where the frame rate is restricted to a few frames per second. Recently there were many advancements in realizing

miniaturized imaging spectrometer for dedicated applications that employ a small number of scans [3]. Such system uses rotational diffraction grating for achieving optical diversity by which information is retrieved.

We aim to build a tunable grating by means of rotation, to achieve optical transfer function diversity. Here we try to overcome the aforementioned problems by making highly efficient polymer diffraction gratings using negative photoresists (SU-8). This diffraction grating can be embedded in MEMS actuator that is fabricated separately based on electrostatic stepper motor principle. The electrostatic motor is designed using flexures and the device works with a move-and-hold mechanism which would be suitable for imaging.

II. MICROFABRICATION

The rotary actuator was fabricated by standard MEMS photolithography techniques. The device was fabricated on a highly conductive silicon-on-insulator (SOI) wafer of $50\,\mu m$ device layer, $3\,\mu m$ oxide layer and $400\,\mu m$ handle layer. The device layer has $< 100 >$ orientation, p-doped with a resistivity of $0.01\,\Omega\,cm$ to $0.1\,\Omega\,cm$. The SOI wafer was initially cleaned with piranha solution at $120\,°C$ for $10\,min$ followed by multiple rounds of rinsing with de-ionized (DI) water. The wafer was treated with hexamethyldisiloxane for $120\,s$ to improve the adhesion between the wafer and resists. The wafer was spin coated with an AZ9260 positive photoresist at $5000\,rpm$ to obtain a uniform thickness of $4.2\,\mu m$. The wafer was hard baked for $4\,min$ at $110\,°C$ and the photolithography was performed using UV light of $365\,nm$. The exposed regions were washed away by 400 metal-ion-free solutions (MIF). For contact pads, the wafer was sputtered with Cr/Au ($20\,nm/400\,nm$) which was followed by lift-off. These contact pads provide electrical connection to the actuator. The process was repeated for device layer patterning of the microactuator. For microactuator, the spin coating was performed using a lower viscosity photoresist, AZ7217 to obtain a thickness of $1.2\,\mu m$ which provides a better resolution in photolithography. After developing, hard baking was performed on a hot plate to make the resist hard followed by DRIE process to etch silicon obtaining nearly vertical walls. The etching was continued till it reaches oxide layer. The microscopic image of the device is shown in Fig. 1(a). The photolithography and DRIE process was repeated on the back side of the wafer to make through hole on the handle layer of SOI wafer. Finally, the sacrificial oxide layer was etched by BOE solution and the wafer was dried using CO_2 critical point dryer.

This work is supported by Singapore Economic Development Board, under the grant number S14-1129-NRF-OSTIn-SRP, and in partly supported by the National Research Foundation Singapore through the Singapore MIT Alliance for Research and Technology (SMART) Centre.

S. S. Muttikulangara, M. Baranski, L. Hu and J. Miao are with the School of Mechanical and Aerospace Engineering, Nanyang Technological University, 50 Nanyang Avenue, Singapore 639798 (email:santhan001@e.ntu.edu.sg, jmiao@pmail.ntu.edu.sg)

S. Rehman is with the Singapore-MIT Alliance for Research and Technology (SMART), 1 CREATE Way Singapore 138602.

978-1-5386-4251-1/18 $31.00 © 2018 IEEE

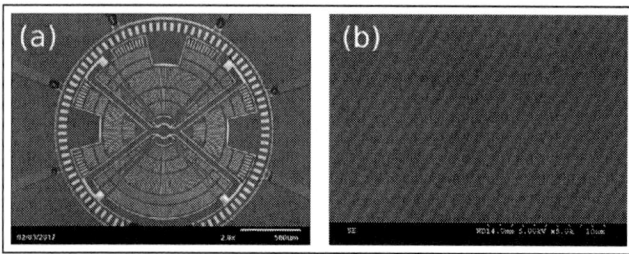

Fig. 1. (a) Microscopic image of the micro fabricated rotary steppermotor before releasing; (b) SEM image of the diffraction grating replicated on top of SU-8 resists.

The diffraction grating was fabricated using replication technique on SU-8 photoresist. SU-8 is a good choice for making replication because of its properties like chemical inertness and mechanical hardness. Replication allows fabrication of blazed angle grating which can achieve high diffraction efficiency. Replication also allows making sub-micron period (Fig. 1(b)) grating which is otherwise difficult to achieve using micromachining technology. This technology provides the flexibility in obtaining the desired grating period for dedicated applications.

III. RESULTS AND DISCUSSIONS

The dispersion spread on the fabricated diffraction grating was experimentally measured on an optical bench. The schematic of the experimental demonstration for measuring dispersion spread of the fabricated grating is shown in Fig. 2(a). A broadband light source (halogen lamp) was focused onto a slit and re-imaged onto a camera plane using 4-F imaging configuration.

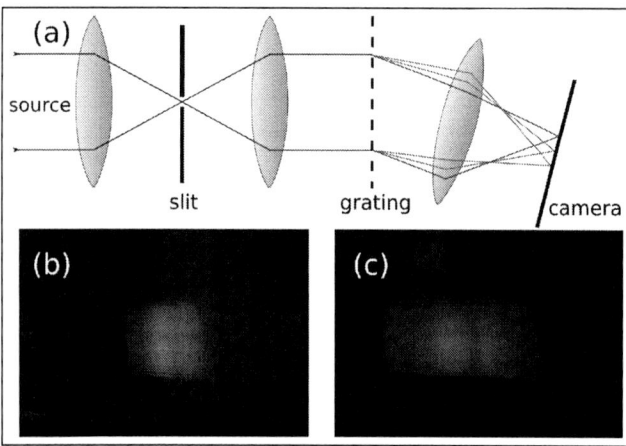

Fig. 2. (a) The schematic of the experimental set-up to measure dispersion spread in the fabricated SU-8 diffraction grating with an illuminated slit, that is re-imaged onto the camera (b) Dispersion due to 600 lines/mm grating showing smaller dispersion spread, and (c) larger dispersion spread when the grating parameter was increased to 1200 lines/mm.

The fabricated diffraction grating was placed on the Fourier plane (2-F) of the 4-F assembly. The diffraction gratings that produce dispersion is manifested on the image plane based on the wavelength of the illumination source. Dispersion spread depends on the grating period. The smaller grating period has large diffraction angle which causes wide dispersion spread. Fig. 2(b) and (c) show the comparison of dispersion spread when two different periods of gratings were used.

The rotary actuator was tested by providing 3-phase half-stepping square voltage waveform with an amplitude of 33 V. The stepping signal resulted in realignment of the movable poles achieving the desired rotation. The plot showing angular displacement as a function of time is depicted in Fig. 3. The actuator was tested by changing the time-delay of the half stepping voltage. The actuator was tested for two step-delay ($t_s = 1$ s and $t_s = 2$ s). This controlled stepping motion allows a precise orientation of grating and can be used with conventional CCD camera for imaging.

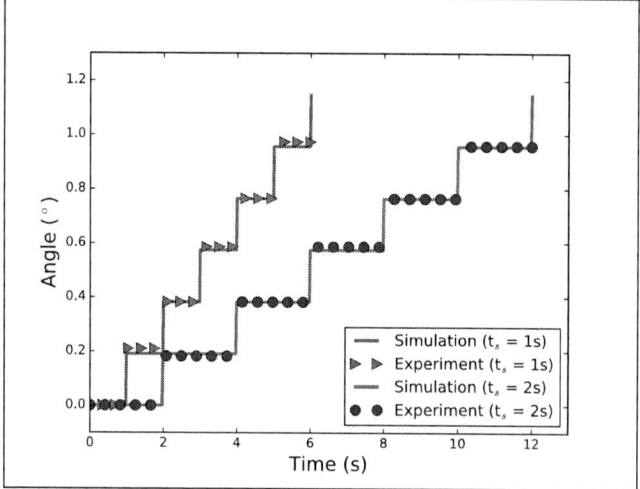

Fig. 3. Experimental plot showing angular change with respect to time steps (t_s) in half stepping voltage.

IV. CONCLUSION

A MEMS-based electrostatic rotary actuator was designed and fabricated using micromachining technology. This actuator is to be used for tuning SU-8 diffraction grating that is fabricated by means of replication technique. Replication facilitated fabrication of blazed angle grating with sub-micron period which is difficult to achieve by conventional photolithography techniques. The device was designed to work using a move-and-hold mechanism which could be suitable for imaging applications.

REFERENCES

[1] J. Miao, R. Lin, L. Chen, Q. Zou, S. Y. Lim, and S. H. Seah, "Design considerations in micromachined silicon microphones," in *International Symposium on Microelectronics and Assembly.* International Society for Optics and Photonics, 2000, pp. 309–314.

[2] S. S. Muttikulangara, M. Baranski, S. Rehman, L. Hu, and J. Miao, "Mems tunable diffraction grating for spaceborne imaging spectroscopic applications," *Sensors*, vol. 17, no. 10, p. 2372, 2017.

[3] M. Baranski, S. Rehman, S. S. Muttikulangara, G. Barbastathis, and J. Miao, "Computational integral field spectroscopy with diverse imaging," *JOSA A*, vol. 34, no. 9, pp. 1711–1719, 2017.

Sensitivity and Q-Factor Trade-off Analysis of MEMS Pressure Sensor for Bladder Implants

Norliana Yusof, *Student Member, IEEE*, Badariah Bais, *Senior Member, IEEE*
Burhanuddin Yeop Majlis, *Senior Member, IEEE*, Norhayati Soin, *Senior Member, IEEE*

Abstract—**Ultra-small and highly sensitive MEMS bladder sensor has to be aligned with efficient wireless coupling to meet the requirements of minimally invasive implantation. In this study, finite element analysis (FEA) is conducted to investigate sensor's sensitivity and Q-factor on MEMS capacitive pressure sensor. The optimum geometry dimension of slotted diaphragm and square coil were analyzed in order to achieve target of high sensor's sensitivity and high quality factor with appropriate frequency sensing meant for bladder sensing application. From capacitive analysis, the optimized parameters were obtained as diaphragm area, A= 1000 um x 1000 um, air gap = 4 um and diaphragm's thickness = 4 um. The capacitive sensitivity of diaphragm is obtained = 2.22 x 10^{-1} mmHg^{-1} and sensor's sensitivity 45.8 kHz/mmHg for operating frequency in range of 55 – 62 MHz. The inductance value of 3.14 uH and Q factor of 63 were obtained at optimum value considering trade off between sensor's sensitivity and the q factor.**

Keywords—*wireless bladder implants; slotted diaphragm, square spiral coil; capacitive sensitivity; Q factor*

I. INTRODUCTION

Wireless sensing for implantable sensor is one of vital criteria as it provides comfort to the patients and reduce infection or complication such as damaged tissue problems created by wires. Selecting suitable operating frequency is important in designing wireless implantable sensor. By operating the sensor at low frequency, it is safer as it reduce the energy absorption of human tissue,but the communication distance will be too limited and also requires large size of the coil [1]. In general, for an implant utilizing radio frequency (RF) communication, a frequency in the low-MHz region provides a good compromise between bandwidth and tissue absorption [2]. The suitable operating frequencies relies in range 10-100 MHz for in-vivo wireless measurements [3].

In this paper, FEA analysis is conducted on geometrical modelling of slotted diaphragm to optimize capacitive sensitivity and suggest suitable operating frequency of sensor in achieving high Q factor and sensor's sensitivity specifically design for inductive coupling an implanted bladder MEMS sensor.

II. FINITE ELEMENT ANALYSIS OF SLOTTED DIAPHRAGM AND SQUARE SPIRAL COIL

Fig. 1 shows the cross sectional view of slotted square diaphragm with square spiral coil and Fig. 2. shows FEA diaphragm model for designed capacitive pressure sensor. The size of MEMS bladder implants should be less than the diameter of human intra-urethral. Hence, the maximum size of outer diameter of the microcoil is should not more than 6 mm after considered the sensor packaging. The square slotted diaphragm was chosen due to its higher induced stress which leads to better sensitivity [4] and ease of fabrication.

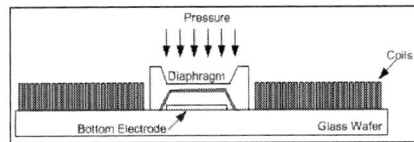

Fig. 1. Cross sectional view of slotted square diaphragm with spiral coil

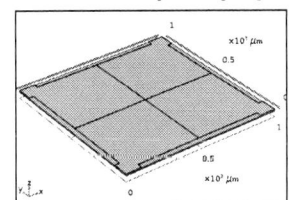

Fig. 2. FEA diaphragm model for designed capacitive pressure sensor

Norliana Yusof is with Institute of Microengineering and Nanoelectronics (IMEN), Universiti Kebangsaan Malaysia, 43600 Bangi, Selangor and Faculty of Innovative Design and Technology, Universiti Sultan Zainal Abidin, Gong Badak Campus, 21300 Kuala Terengganu, Malaysia (e-mail:P86120@ siswa.ukm.edu.my).

Burhanuddin Yeop Majlis is with Institute of Microengineering and Nanoelectronics (IMEN), Universiti Kebangsaan Malaysia, 43600 Bangi, Selangor,Malaysia.

Badariah Bais is with Department of Electrical Electronics and Systems Engineering, Faculty of Engineering and Built Environment and Institute of Microengineering and Nanoelectronics (IMEN), Universiti Kebangsaan Malaysia, 43600 Bangi, Selangor,Malaysia.

Norhayati Soin is with Department of Electrical Engineering, Faculty of Engineering, University of Malaya, 50603 Kuala Lumpur, Malaysia.

TABLE I.   Q FACTOR AND SENSOR SENSITIVITY OF SPIRAL COIL

Number of turns	Outer Diameter (μm)	Induc-tance (μH)	Frequency Response (MHz)	Sensor Sensi-tivity (kHz/mm Hg)	Q Factor
25	3030	1.56	77.6 - 88.0	64.9	50.7
30	3430	2.34	63.3 - 71.8	53.0	57.8
35	3830	3.34	53.0 - 60.1	44.4	64.7
40	4230	4.56	45.4 - 51.4	38.0	71.5
45	4630	6.04	39.4 – 44.7	33.0	78.2
50	5030	7.80	34.7 – 39.3	29.0	84.8

III.   RESULTS AND DISCUSSION

A. Capacitive Sensitivity Analysis of Slotted Diaphragm

Fig. 3 illustrates the capacitive sensitivity of 4, 5 and 6 μm diaphragm thickness with varied air gap in between 4-10 μm. The maximum capacitive sensitivity can be seen at 4μm diaphragm thickness and air gap under 0-160 mmHg applied pressure. Hence, the optimized parameters; diaphragm area, A = 1000 um x 1000 um, air gap = 4 um and diaphragm's thickness = 4 um is used for designing diaphragm and the capacitance changes in between 2.11pF to 2.70 pF is applied for further analysis.

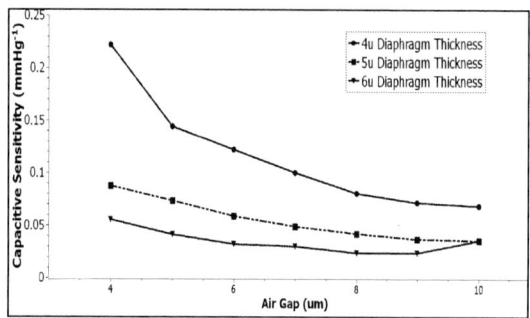

Fig. 3. Air gap versus capacitive sensitivity for different diaphragm thickness

B. Q Factor and Inductance Analysis of Spiral Coil

Table I shows the Q factor of the coil and sensor sensitivity with changes of the coil's dimension. The square spiral coil is fixed at 30 μm height and width with 10 μm of coil gap. The copper coil is varied by number of turns in between 25 to 50 turns. From Table I, the sensor sensitivity decreased and the Q factor increased as number of turns increased. This is the trade off that should be considered in designing the wireless MEMS pressure sensor. Fig. 4 illustrates the Q factor and sensor sensitivity is compensated at 34 turns and this gives the inductance value of 3.14 μm. The change in resonant frequency of the sensor under 0-160 mmHg is shown in Fig.5. From Fig.5, the resonant frequency (55-62MHz) is linearly changes subjected to applied pressure with 45.8 kHz/mmHg sensor sensitivity and Q factor of the coil is 63.

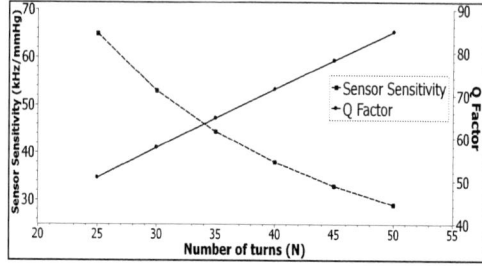

Fig. 4. Trade off between sensor sensitivity and Q factor

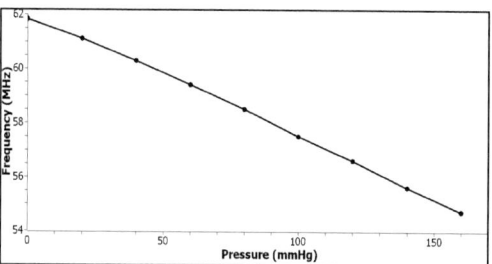

Fig. 5. Resonant frequency with applied pressure

IV.   CONCLUSION

Analysis of LC wireless MEMS capacitive pressure sensor were designed and presented using COMSOL Multiphysics. The capacitive sensitivity analysis was conducted by COMSOL software shows the capacitance changes in between 2.11pF to 2.70 pF at its optimum diaphragm parameter. The optimum inductance value was analysed after considering the trade off between sensor sensitivity and coil Q factor. From this study, it can be concluded that the parameters suggested is suitable choice for designing the implantable MEMS bladder pressure sensor due to its high sensitivity and Q factor and meet the requirement of desirable operating frequency range and size for the implants bladder sensor. However, the determination of geometric parameters need to consider the process technology limits and facilities constraint to realize the fabricated structure.

REFERENCES

[1]   J. Zhai, T. V. How, and B. Hon, "Design and modelling of a passive wireless pressure sensor," *CIRP Ann. - Manuf. Technol.*, vol. 59, no. 1, pp. 187–190, 2010.

[2]   J. A. Potkay, "Long term, implantable blood pressure monitoring systems," *Biomed. Microdevices*, vol. 10, no. 3, pp. 379–392, 2008.

[3]   M. Luo, A. W. Martinez, C. Song, F. Herrault, and M. G. Allen, "A microfabricated wireless RF pressure sensor made completely of biodegradable materials," *J. Microelectromechanical Syst.*, vol. 23, no. 1, pp. 4–13, 2014.

[4]   N. Yusof, B. Bais, and B. Y. Majlis, "Mechanical Analysis of MEMS Diaphragm for Bladder Pressure Monitoring," Presented in *2017   IEEE Regional Symposium on Micro and Nanoelectronics - Proc.*, pp. 22-25, 2017. To be published.

Effect of Fluorine Circumference of Zinc-hexadecafluorophthalocyanine towards VOCs determination by Using Low-Cost Optical Electronic Nose

Treenet Thepudom[1] and Teerakiat Kerdcharoen[2,*]

[1]*Materials Science and Engineering Program, Faculty of Science, Mahidol University, Bangkok, Thailand.*

[2]*Department of Physics, Faculty of Science, Mahidol University, Bangkok, Thailand*

[3]*NANOTEC Center of Excellence at Mahidol University, National Nanotechnology Center Bangkok, Thailand*

*Corresponding author: teerakiat@yahoo.com

Abstract—**Organic thin film gas sensor based on Zinc hexadecafluorophthalocyanine (ZnPcF$_{16}$) have been fabricated by spin-coating method. The thin film was then used to study the absorption spectra and morphology using UV-VIS spectrometer. Then the fabricated film has been used as a chemical sensing material for the detection of VOCs (ammonia, acetone, ethanol and methanol) by using a low-cost optical measurement. For getting more insight, Quantum mechanical calculation based on DFT was employed to investigate the interactions between VOCs and sensing molecules.**

I. INTRODUCTION

Metal phthalocyanines (MPc) are interesting molecules with broad ranges of potential industrial applications such as solar cell and chemical sensor [1]. These organic pigments are widely used as sensing material due to their outstanding ability of diversified substitution of central atom by various types of metal atom such as Fe, Co, Ni, Cu, Zn and Mg, which interact differently with some oxidising and reducing gases as well as volatile organic compounds (VOCs) by both physical and chemical adsorption mechanism. Zinc phthalocyanine (ZnPc) is one of the most popular metal phthalocyanine compounds with a significant number of researchers studying the gas-sensing properties and doing fabrication of thin film gas sensors for detection of various VOCs [2, 3]. Among the ZnPc compounds, Zinc hexadecafluorophthalocyanine (ZnPcF$_{16}$) is quite unique, of which all the hydrogen atoms at the circumference are substituted by the fluorine atoms with high electro-negativity. However, only a few papers have been published about ZnPcF$_{16}$ sensitivity and gas sensing properties [4, 5]. The uniqueness of this compound presents both potential and challenge to develop as a gas-sensing material.

In this work, we have studied the gas-sensing properties of ZnPcF$_{16}$ with alcohol vapors (acetone, ethanol and methanol) and ammonia based on a combined theoretical-experimental approach. It is hoped that the understanding obtained from this study will assist in the development of novel gas sensors for future applications in the food and environment areas.

II. EXPERIMENTAL APPROACHES

A. Computational method

To investigate the sensing mechanism of ZnPcF$_{16}$ with analyte molecules: ethanol and methanol, Density functional theory (DFT) has been usefully applied to predict the structural and electronic properties of the sensing molecule. Such method was first started by geometry optimization of all molecules using semi-empirical PM3 method following the DFT method at basis set of B3LYP/6-31G*. In such study, the interactions between sensing materials and VOCs were investigated by varying the distance between the zinc atom of the sensing molecules (ZnPcF$_{16}$) and the oxygen atom of each alcohol molecules, nitrogen atom of ammonia.

Finally, the interaction energy between sensing and gas molecule was calculated by equation:

$$E_{Int} = [E_{MP+VOC} - (E_{MP} + E_{VOC})] \qquad (1)$$

where E_{Int} is the interaction energy between a sensing molecule and a VOC molecule. E_{MP+VOC} denotes the total energy of the interacting pair. E_{MP} and E_{VOC} represent the total energy of the sensing and VOC molecules, respectively.

B. Experimental method

To fabricate thin film gas sensor, ZnPcF$_{16}$ was dissolved in tetrahydrofuran solution at a concentration of 3 mg/ml. The solution was then spin-coated on glass substrate at a spin speed of 1200 rpm for 20 seconds and then dried at 80°C. The gas sensing properties of the thin film with VOCs were investigated by measuring the absorption change of the thin film upon exposure to the VOCs at room temperature (the experimental set up is illustrated in Fig.1).

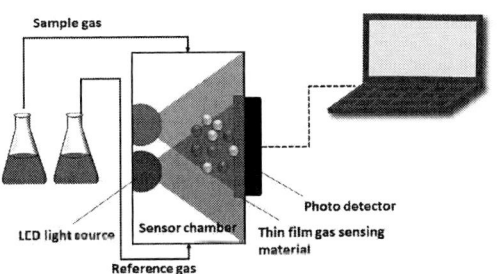

Figure 1. Schematic diagram of VOCs detection using a low-cost optical measurement.

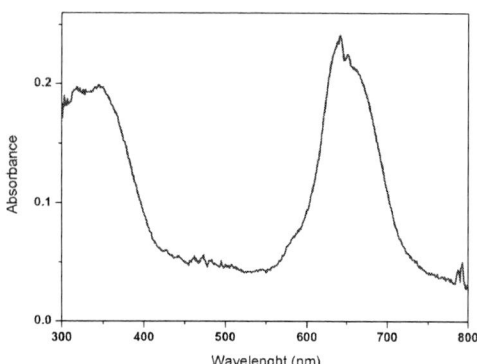

Figure 2. UV-Vis spectrum of ZnPcF$_{16}$ spin-coated film.

III. RESULT AND DISCUSSION

A. Characterization of ZnPcF$_{16}$

ZnPcF$_{16}$ thin film was investigated on their absorption spectra by using Jenway UV-VIS spectrometer in the range of 300-800 nm at room temperature. Fig. 2 shows absorption spectra of ZnPcF$_{16}$ spin-coated at the spin speeds of 1200 rpm. The intensive Q-band at 640 nm is attributed to $\pi \rightarrow \pi^*$ transitions from the HOMO (highest occupied molecular orbital) to LUMO (lowest unoccupied molecular orbital), while the soret (B)-band is observed at 340 nm.

B. Gas sensing ability of ZnPcF$_{16}$ thin film

The electronic structure calculation based on B3LYP/6-31G* was employed to investigate the interaction energies between sensing molecule to analytes molecule. Fig. 3 shows the plots of the interaction energies of the sensing molecules to those analytes molecule. The results reveal that ZnPcF$_{16}$ has a stronger interaction with ammonia than acetone, ethanol and methanol, respectively. The lowest interaction energy was carried out at the optimized distance of 2.8 angstrom with a total interaction energy of -9.92 kcal/mol. Gas sensing ability of ZnPcF$_{16}$ with ammonia, acetone, ethanol and methanol vapors were also investigated by the optical measurement.

Fig. 4 shows the sensing response of ZnPcF$_{16}$ film to target gases under violet and red light sources (appropriate wavelength for the absorption peaks of sensing material is

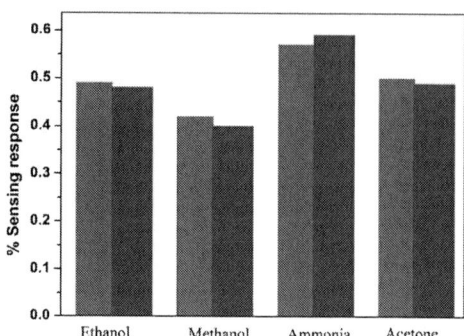

Figure 4. Sensing response plots of ZnPcF$_{16}$ thin film with analytes vapor at 200 ppm of concentration under red and violet LED lights.

340 and 640 nm, respectively). The percent changes of this measurement illustrate that ZnPcF$_{16}$ film has an ability to become a gas sensing material for those tested gases, especially ammonia. Furthermore, the experimental results are in agreement with the theoretical results, in which the most interacting pair of analyte-sensor molecules, e.g., ammonia, yield best sensing signals.

CONCLUSION

In this work, the sensing ability of ZnPcF$_{16}$ film toward VOCs, such as ethanol, methanol, acetone and ammonia, was investigated. The experimental results as compared to the density functional theory (DFT) calculations confirm that ZnPcF$_{16}$ could be performed as gas sensor for such gases. Thus, the ZnPcF$_{16}$ spin-coated film together with low-cost LED-based electronic nose will be applied for food or environmental monitoring in future works.

ACKNOWLEDGMENT

This research was supported by Mahidol University, National Nanotechnology Center, and Research and Researchers for Industries (RRI) scholarship of the Thailand Research Fund.

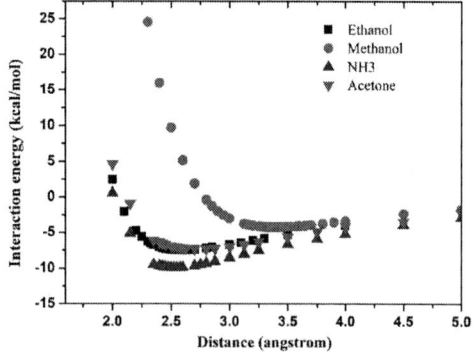

Figure 3. Interaction energy plots with different atomic distances between sensing and analytes molecule.

REFERENCES

[1] M. G. Walter, A. B. Rudine and C. C. Wamser, "Porphyrins and phthalocyanines in solar photovoltaic cells," *Porphyrins Phthalocyanines*, 2010, vol. 14, pp. 759.

[2] S. Kladsomboon and T. Kerdcharoen, "A Method for Detection of Alcohol Vapors Based on Optical Sensing of MgTPP Thin Film by Optical Spectrometer and Principal Component Analysis," *Analytica Chimica Acta*, 2012, vol. 757, pp. 75-82.

[3] T. Thepudom, S. Kladsomboon, T. Pogfay, A. Tuantranont and T. Kerdcharoen, "Portable optical-based electronic nose using dual-sensors array applied for volatile discrimination," *Electrical Engineering/Electronic Computer Telecommunication and Information Technology conference (ECTI2012)*, 2012, pp. 1-4.

[4] B. Schollhorna, J. P. Gernainb, A. Paulyb, C. Maleyssonb and J. P. Blancb, "Influence of peripheral electron-withdrawing substituents on the conductivity of zinc phthalocyanine in the presence of gases. Part 1: reducing gases," *Thin Solid Film*, 1998, vol. 326, pp. 245-250.

[5] T. Ikame, K. Kanai, Y. Ouchi, E. Ito, A. Fujimori and K. Seki, "Molecular orientation of F$_{16}$ZnPc deposited on Au and Mg substrates studied by NEXAFS and IRRAS," *Chemical Physics Letters*, 2005, vol. 413, pp. 373-378.

Analytical Capacitance Model for In-Ga-Zn-O Thin-Film Transistors Including Degeneration

Feng Zhuang , Jielin Fang, Wanling Deng, Xiaoyu Ma, and Junkai Huang

Abstract—**In this paper, a capacitance model of amorphous In-Zn-Ga-O thin-film transistors is proposed based on terms of surface potential. The carrier degeneracy is considered due to the device characteristics. Based on the symmetric quadrature version of Charge-Sheet Model, a new expression of terminal charge is provided. Therefore, the capacitance model can be obtained. The validity of the model is verified by comparisons with measured data and simulation results. It is very useful to circuit simulators.**

I. INTRODUCTION

Amorphous indium gallium zinc oxide(a-IGZO), due to the excellent performances such as high mobility, low-cost and room-temperature fabrication processes, optical transparency, low leak current and large-area uniform integration on flexible substrates, has been recognized as the most promising candidate for the next TFTs technology. In the near future, a-IGZO TFTs have been already found applications in many fields. Therefore, besides maintaining high computational efficiency, a good compact model is required to accurately capture all real-device physical effects. Due to the strong electronegativity, the conduction band of a-IGZO is composed of spherical overlapping s orbitals from In ionic and the valence band is composed of 2p orbitals from O ionic[1], leading to a smaller density of band tail states than that of covalent semiconductors like hydrogenated amorphous silicon (a-Si)[2]. Meanwhile, the existence of a high density of deep trap states near the valence band maximum (E_V)[3] pins the Fermi level and stops it to move toward E_V. So at present, the a-IGZO material is targeted for n-type TFTs. The band structure and trap states distribution of IGZO TFTs illustrate that the Fermi level may exceed the conduction band minimum (E_C) at a large gate voltage. It makes the electrons to transport in the non-localized band states. In other words, the degenerate conduction regime must be taken into account in the compact models. In the last years, there are few TFT models focus on the degenerate mechanism. M. Ghittorelli [4,5] at the first time proposed an analytical model that described the charge carrier concentration valid for both non-degenerate and degenerate conductions. But they did not explicitly solve the surface potential which is very essential since the drain current or other quantities are most sensitive to the surface potential. The Symmetric Quadrature Method (SQM) was proposed by L. Colalongo[6,7], it is useful to our work. SQM can be

viewed as a generalization to TFTs modeling of the silicon MOSFET models based on the symmetric linearization method (SLM) such as SP[8] and PSP[9]. This approach eliminates the need of a trade-off between the accuracy and the simplicity of the current and charge formulations.

In this paper, based on J. L. Fang et al.[10], we propose a unified capacitance compact model for IGZO TFTs by taking the advantages of the SQM. Both non-degenerate and degenerate conductions are included in compact model.

II. CAPACITANCE MODEL

We assumed an n-type and un-doped a-IGZO TFT with tail states. Fig. 1(a) shows its schematic diagram of the structure, and Fig. 1(b) shows the cross section of the components of the extrinsic parasitic capacitance. $C1,C2,C3$ which are the outer edge field capacitance, coverage capacitance and inner edge field capacitance between the gate and source and drain, respectively.

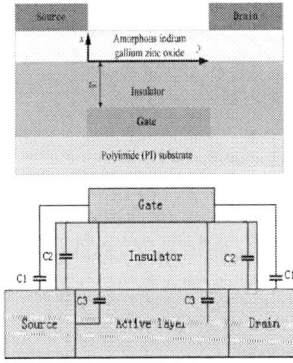

Figure 1. (a)Diagram of the structure of an a-IGZO TFT. (b) Cross section of the components of extrinsic parasitic capacitance.

Similar to the DC model, the dynamic model is also important to the circuit simulation tools. Based on the results of φ_s[10], the dynamic behavior of the terminal charges in the degenerate states is proposed by using the symmetric quadrature method(SQM). For the gate charge, it is given by:

$$Q_G = WC_{OX}R\int_{-\Delta s/2}^{\Delta s/2}(V_{gb}-s-\varphi_{sm})(Z_1+Z_2s+\beta s^2)ds. \quad (1)$$

According to the Ward-Dutton charge partitioning[11], and ignoring the carrier recombination effect, the source and drain charges are given by:

$$Q_S = W\int_0^L(1-\frac{y}{L})Q_a dy. \quad (2)$$

$$Q_D = W\int_0^L\frac{y}{L}Q_a dy. \quad (3)$$

*This work was supported in part by the Fundamental Research Funds for the Central Universities under Grant 21617405.

F. Zhuang, J. Fang, W. Deng, X. Ma and J. Huang are with the Department of Electronic Engineering, Jinan University, Guangzhou 510630, China (corresponding author: W. Deng, e-mail: dwanl@126.com).

where Q_a is the free carrier concentration per unit area at the semiconductor/oxide interface and equals[5] $Q_a=Q_{am}+\alpha s+\beta s^2$, $s=\varphi_s-\varphi_{sm}$, Q_{am} is the value of Q_a at the position of mid-potential($\varphi_{sm} = (\varphi_{ss}+\varphi_{sd})/2$), α and β are the coefficients of the quadratic function. Herein $R=W\mu_{eff}/I_{ds}$, $Z_1=Q_{am}-\alpha\phi_t$, $Z_2=\alpha-2\beta\phi_t$.

The total accumulation charge in the channel is given by

$$Q_A = Q_S + Q_D = W\int_0^L Q_a dy$$
$$= WR\int (Q_{am} + \alpha s + \beta s^2)(Z_1 + Z_2 s + \beta s^2)ds. \tag{4}$$

Equation (4) can be rewritten as

$$Q_A = WR\int_{-\Delta s/2}^{\Delta s/2} T_0 + T_1 s + T_2 s^2 + T_3 s^3 + T_4 s^4 ds. \tag{5}$$

where $T_0=Q_{am}Z_1$; $T_1=Q_{am}Z_2+\alpha Z_1$; $T_2=Q_{am}\beta+\beta Z_1+\alpha Z_2$; $T_3=\beta Z_2+\alpha\beta$; and $T4=\beta^2$. Solving the integral yields.

$$Q_A = WR(T_0\Delta s + \frac{T_2}{12}\Delta s^3 + \frac{T_4}{80}\Delta s^5). \tag{6}$$

The corresponding Q_S, Q_D can be expressed as:

$$Q_S = \frac{WR}{L}\left(T_{s1}\Delta s + \frac{T_{s2}}{12}\Delta s^3 + \frac{T_{s3}}{80}\Delta s^5 + \frac{T_{s4}}{448}\Delta s^7\right). \tag{7}$$

$$Q_D = \frac{WR}{L}\left(T_{d1}\Delta s + \frac{T_{d2}}{12}\Delta s^3 + \frac{T_{d3}}{80}\Delta s^5 + \frac{T_{d4}}{448}\Delta s^7\right). \tag{8}$$

where

$$T_{s1} = LT_0 - y_m T_0;$$
$$T_{s2} = LT_2 - RZ_1 T_1 - RZ_2 T_1 / 2 - y_m T_2;$$
$$T_{s3} = LT_4 - y_m T_4 - RZ_1 T_3 - RZ_2 T_2 / 2 - \beta RT_1 / 3;$$
$$T_{s4} = -RZ_2 T_4 / 2 - \beta RT_3 / 3;$$
$$T_{d1} = y_m T_0;$$
$$T_{d2} = RZ_1 T_1 + RZ_2 T_1 / 2 + y_m T_2;$$
$$T_{d3} = y_m T_4 + RZ_1 T_3 + RZ_2 T_2 / 2 + \beta RT_1 / 3;$$
$$T_{d4} = RZ_2 T_4 / 2 + \beta RT_3 / 3.$$

As shown in Fig. 2, at $V_{gs}=20V$, the model results of the charges Q_A, Q_S, Q_D as function of V_{ds} are compared with the numerical solution.

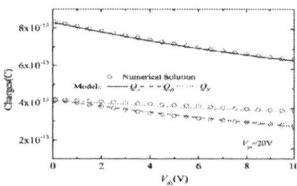

Figure 2. The model results of the charges Q_A, Q_S, Q_D as function of V_{ds} compared with the numerical solution.

Using the results of teminal charges, we can obtain the intrinsic capacitances (C_{ij}) of a-IGZO TFTs. As shown in Fig. 3, the proposed model gives a reasonably accurate description in a wide range of bias conditions[12] and the model results of normalized C_{gs}, C_{sg}, C_{gd} and C_{dg} as a function of V_{gs}.

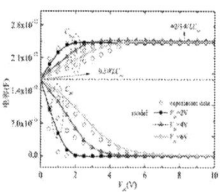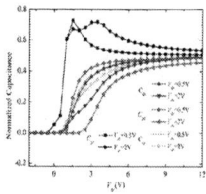

Figure 3. (a)Comparison of C_{gs}-V_{gs} and C_{gd}-V_{gs} characteristics between model result and experimental data. (b) The model results of normalized capacitances C_{gs}, C_{sg}, C_{gd} and C_{dg} as a function of V_{gs}.

III. CONCLUSION

In this article, extending the SQM method, we develop a capacitance compact model by using a simple algorithm and clear physical conception. Furthermore the validity of the model is supported by the verification results, which is suitable for implementation in EDA simulators.

REFERENCES

[1] T. Kamiya, K. Nomura, H. Hosono, "Origins of High Mobility and Low Operation Voltage of Amorphous Oxide TFTs: Electronic Structure, Electron Transport, Defects and Doping" *Journal of Display Technology*, vol. 5, no. 12, pp. 468-483, Nov. 2009.

[2] H. H. Hsieh, T. Kamiya, K. Nomura, et al. "Modeling of amorphous InGaZnO4 thin film transistors and their subgap density of states," *Applied Physics Letters*, vol. 92, no. 13, pp. 133503, Mar. 2008.

[3] K. Nomura, T. Kamiya, H. Yanagi, et al. "Subgap states in transparent amorphous oxide semiconductor, In-Ga-Zn-O, observed by bulk sensitive x-ray photoelectron spectroscopy," *Applied Physics Letters*, vol. 92, no. 20, pp. 202117, May. 2008.

[4] M. Ghittorelli, F. Torricelli, Z. M. Kovács-Vajna, "Analytical physical-based drain-current model of amorphous InGaZnO TFTs accounting for both non-degenerate and degenerate conduction," *IEEE Electron Device Letters*, vol. 36, no. 12, pp. 1340-1343, Dec. 2015.

[5] M. Ghittorelli, F. Torricelli, Z. M. Kovács-Vajna, "Physical modeling of amorphous InGaZnO thin-film transistors: The role of degenerate conduction," *IEEE Trans. Electron Devices*, vol. 63, no. 6, pp. 2417-2423, Jun. 2016.

[6] L. Colalongo, "Compact model of amorphous InGaZnO thin film transistors based on symmetric quadrature of accumulation charge," *IEEE Electron Device Letters*, vol. 37, no. 4, pp. 416-418, Apr. 2016.

[7] L. Colalongo, "DC/Dynamic Surface Potential Based Model of InGaZnO Transistors for Circuit Simulation," *Journal of Display Technology*, vol. 12, no. 12, pp. 1514-1521, Dec. 2016.

[8] G. Gildenblat, et al. "SP: an advanced surface-potential-based compact MOSFET model," *Custom Integrated Circuits Conference*, 2003. *Proceedings of the IEEE. IEEE Xplore*, pp.233-240.

[9] R. V. Langevelde, G. Gildenblat, "PSP: An advanced surface potential based MOSFET model," *IEEE Trans. Electron Devices*, vol. 53, no. 9, pp. 1979-1993, Aug. 2016.

[10] J. L. Fang, W. L. Deng, X. Y. Ma, J. K. Huang, "A Surface potential based DC Model of Amorphous Oxide Semiconductor TFTs Including Degenvoleration," *IEEE Electron Devices Letters*, vol. 38 no. 2, pp. 183-186, Dec. 2016.

[11] D. E. Ward and R. W. Dutton, "A charge-oriented model for MOS transistor capacitances," *IEEE Journal of Solid-State Circuits*, vol. 13, no. 5, pp. 703-708, Oct. 1978.

[12] P.G. Bahubalindruni, V.G. Tavares, P. Barquinha, R. Martins, et al, "InGaZnO TFT behavioral model for IC design," *Analog Integrated Circuits and Signal Processing*, vol. 87, no. 1, pp. 1-8, Apr. 2016.

AUTHOR INDEX

Baek, C.-K. 3
Bais, B. 49
Baranski, M. 47
Bodelot, L. 7
Chen, L. 45
Chen, Y. 29
Chi, C. 41
Deng, W. 53
Fang, J. 53
Giorcelli, M. 33
Gu, H. 13
Hatta, S. 11, 19
Hu, L. 47
Huang, J. 53
Ibrahim, F. 25, 27
Iliescu, C. 17
Jaafar, A. 11
Kanhere, E. 43
Kerdcharoen, T. 5, 51
Kong, Q. 7
Lebental, B. 7
Lee, T. 21
Li, T. 41
Liu, C.-K. 39, 41
Liu, X. 13
Ma, X. 53
Majlis, B. 49
Marculescu, C. 17
Miao, J. 9, 43, 47
Misra, D. 7
Mohamad, M. 25
Muttikulangara, S. 47
Ni, M. 17
Ong, B. 35
Ong, S. 31
Othman, N. 19
Pobkrut, T. 5
Raghavan, N. 15
Rahman, S. 27
Rehman, S. 47
Ren, S. 23, 37, 45
Savi, P. 33
Shen, Q. 23, 37, 45
Siyang, S. 5
Soin, N. 11, 19, 25, 27, 49
Sun, J. 39
Sun, X. 41
Tagliaferro, A. 33
Tan, C. 1
Tang, L. 9, 29
Tao, K. 9, 29, 43
Tay, B. 7
Thepudom, T. 5, 51
Tok, E. 31, 35
Tresset, G. 17
Triantafyllou, M. 43
Wang, B. 29
Wang, C. 39
Wang, F. 37
Wang, N. 43
Wu, J. 9, 43
Xie, H. 45
Xie, J. 23, 37, 45
Xue, N. 39, 41
Yang, J. 23
Yoon, J.-S. 3
Yuan, W. 23, 37, 45
Yusof, N. 49
Zhang, J. 37
Zhao, B. 13
Zhou, H. 13
Zhuang, F. 53